T0234103

Robust Control of DC-DC Converters

The Kharitonov's Theorem Approach with MATLAB® Codes

Synthesis Lectures on Power Electronics

Editor
Jerry Hudgins, *University of Nebraska*

Synthesis Lectures on Power Electronics will publish 50- to 100-page publications on topics related to power electronics, ancillary components, packaging and integration, electric machines and their drive systems, as well as related subjects such as EMI and power quality. Each lecture develops a particular topic with the requisite introductory material and progresses to more advanced subject matter such that a comprehensive body of knowledge is encompassed. Simulation and modeling techniques and examples are included where applicable. The authors selected to write the lectures are leading experts on each subject who have extensive backgrounds in the theory, design, and implementation of power electronics, and electric machines and drives.

The series is designed to meet the demands of modern engineers, technologists, and engineering managers who face the increased electrification and proliferation of power processing systems into all aspects of electrical engineering applications and must learn to design, incorporate, or maintain these systems.

Transient Electro-Thermal Modeling of Bipolar Power Semiconductor Devices
Tanya Kirilova Gachovska, Bin Du, Jerry L. Hudgins, and Enrico Santi
2013

Modeling Bipolar Power Semiconductor Devices
Tanya K. Gachovska, Jerry L. Hudgins, Enrico Santi, Angus Bryant, and Patrick R. Palmer
2013

Signal Processing for Solar Array Monitoring, Fault Detection, and Optimization
Mahesh Banavar, Henry Braun, Santoshi Tejasri Buddha, Venkatachalam Krishnan, Andreas
Spanias, Shinichi Takada, Toru Takehara, Cihan Tepedelenlioglu, and Ted Yeider
2012

The Smart Grid: Adapting the Power System to New Challenges
Math H.J. Bollen
2011

Digital Control in Power Electronics
Simone Buso and Paolo Mattavelli
2006

Power Electronics for Modern Wind Turbines
Frede Blaabjerg and Zhe Chen
2006

Robust Control of DC-DC Converters: The Kharitonov's Theorem Approach with MATLAB® Codes
Farzin Asadi

ISBN: 978-3-031-01375-1   paperback
ISBN: 978-3-031-02503-7   ebook
ISBN: 978-3-031-00324-0   hardcover

DOI 10.1007/978-3-031-02503-7

A Publication in the Springer series
SYNTHESIS LECTURES ON POWER ELECTRONICS

Lecture #11
Series ISSN
Print 1931-9525   Electronic 1931-9533

Robust Control of DC-DC Converters

The Kharitonov's Theorem Approach
with MATLAB® Codes

Farzin Asadi
Kocaeli University, Kocaeli, Turkey

SYNTHESIS LECTURES ON POWER ELECTRONICS #11

ABSTRACT

DC-DC converters require negative feedback to provide a suitable output voltage or current for the load. Obtaining a stable output voltage or current in the presence of disturbances like input voltage changes and/or output load changes seems impossible without some form of control.

This book shows how simple controllers such as Proportional-Integral (PI) can turn into a robust controller by correct selection of its parameters. Kharitonov's theorem is an important tool toward this end.

This book consist of two parts. The first part shows how one can obtain the interval plant model of a DC-DC converter. The second part introduces the Kharitonov's theorem. Kharitonov's theorem is an analysis tool rather than a design tool. Some case studies show how it can be used as a design tool.

The prerequisite for reading this book is a first course on feedback control theory and power electronics.

KEYWORDS

control of DC-DC converters, dynamics of DC-DC converters, interval polynomial, interval plant, Kharitonov's theorem, modeling of power electronics converters, robust control, 16-plant theorem, state space averaging

I dedicate this book to my parents and my lovely family.

I appreciate the author's effort to convert my theorem into a useful tool for engineers.

I am glad that my paper still attracts the attention of both scholars and practicing engineers.

Prof. Vladimir Kharitonov
Saint Petersburg State University

Contents

Preface

Physical systems cannot be described exactly by a mathematical model. A model, no matter how detailed, is never a completely accurate representation of a real physical system.

Robust control is a design methodology that explicitly deals with uncertainty. Robust control designs a controller such that:

- some level of performance of the controlled system is guaranteed and

- irrespective of the changes in the plant/process dynamics within a predefined class, the stability is guaranteed.

Robust Controller Design can be done using techniques such as: H_∞, μ synthesis, Kharitonov's theorem, Linear Matrix Inequality (LMI), and Quantitative Feedback Theory (QFT), to name a few. Kharitonov's theorem-based robust controller design is much easier than other techniques. Successful application of techniques such as H_∞, μ, LMI, or QFT requires advanced mathematics and considerable education. On the other hand, Kharitonov's theorem-based methods require only algebraic manipulation and basic computer programming skills.

This book shows how simple controllers such as Proportional-Integral (PI) can turn into a robust controller by correct selection of its parameters. Kharitonov's theorem is an important tool toward this end.

It is assumed that the reader knows the basics of DC-DC converters and linear control theory. There are plenty of textbooks available on power electronics and linear control theory. You can refer to the given references Ang and Oliva [2005]–Rashid [2013] if you need to review the concepts.

The book is summarized as follows.

Chapter 1 shows how one can extract the uncertain model of DC-DC converters. Kharitonov's theorem is used to design the robust controller based on the obtained model.

Chapter 2 introduces the Kharitonov's theorem. Kharitonov's theorem is an analysis tool rather than a design tool. This chapter shows how it can be used as a design tool.

Chapter 3 shows the design procedure with some case studies.

I hope that this book will be useful to the readers, and I welcome comments on the book.

Farzin Asadi
farzin.asadi@kocaeli.edu.tr
August 2018

REFERENCES

Ang, S. and Oliva, A. *Power Switching Converters*, Taylor & Francis, 2005. xiii

Erikson, R. and Maksimovic, D. *Fundamentals of Power Electronics*, Springer, 2001. DOI: 10.1007/b100747.

Franklin, G. F., Powel, J.D., and Emami-Naeini, A. *Feedback Control of Dynamic Systems*, Prentice Hall, 2002.

Hart, D. *Power Electronics*, McGraw Hill, 2011.

Kazimierczuck, M. K. *Pulse Width Modulated DC-DC Power Converters*, John Wiley, 2012.

Mohan, N. and Undeland, T. *Power Electronics Converters, Applications and Design*, Wiley, 2002.

Nise, N. S. *Control Systems Engineering*, John Wiley and Sons, 2011.

Ogata, K. *Modern Control Engineering*, Pearson, 2009.

Rashid, M. *Power Electronics Devices, Circuits and Applications*, Pearson, 2013. xiii

Acknowledgments

The author gratefully acknowledges the MathWorks® support for this project.

Farzin Asadi
August 2018

CHAPTER 1

Extraction of Uncertain Model of DC-DC Converters

1.1 INTRODUCTION

A switched DC-DC converter takes the voltage from a DC source and converts the voltage of supply into another DC voltage level. It is used to increase or decrease the input voltage level.

Switched DC-DC converters have become an essential component of industrial applications in recent decades. Their high efficiency, small size, low weight, and reduced cost make them a good alternative for conventional linear power supplies, even at low power levels.

Switched DC-DC converters are nonlinear variable structure systems. Various techniques can be found in literature to obtain a Linear Time Invariant (LTI) model of a switched DC-DC converter. The most well-known methods are: current injected approach [Kislovski et al., 1991], circuit averaging [Erikson and Maksimovic, 2007], and State Space Averaging (SSA) [Middlebrook and Cuk, 1977]. A comprehensive survey of the modeling issues can be found from Maksimovic et al. [2001].

Dynamics of switched DC-DC converters change under different output load conditions and/or input voltage. Component tolerances affect the converter dynamics as well. So, a switched DC-DC converter can be modeled as an uncertain dynamical system. Robust control techniques can be used to design controllers for such systems. Robust control refers to control synthesis, its performance, and stability, not only to the nominal plant model, but also to the whole family of models in the area of permitted uncertainty of the modeling (perturbations are bounded). So, new terms appear such as robust performance (guaranteed performance of the control system for all systems in the uncertainty region) and robust stability (stability of all possible systems inside the uncertainty modeling bounds).

A number of robust control techniques have been applied successfully to DC-DC converters. Obtaining the uncertain model of the converter is an important step in designing a robust controller. This chapter shows how the uncertainties model of a DC-DC converter can be extracted.

This chapter studies a buck converter as an illustrative example. However, the studied techniques are general and can be applied to other types of switching DC-DC converters as well.

This chapter is organized as follows. Different types of uncertainties are studied in the second section. The third section reviews some of the important robust controller design techniques applied to DC-DC converters. Dynamics of a buck converter without uncertainty is studied in the fourth section. Effect of changes in component values and operating conditions on the converter dynamics is studied in the fifth section.

1.2   UNCERTAINTY MODELS

Modeling is the process of formulating a mathematical description of the system. A model, no matter how detailed, is never a completely accurate representation of a real physical system. A mathematical model is always just an approximation of the true, physical reality of the system dynamics.

Uncertainty refers to the differences or errors between model and real systems and whatever methodology is used to present these errors will be called an uncertainty model. Successful robust control-system design would depend on, to a certain extent, an appropriate description of the perturbation considered.

1.2.1   PARAMETRIC UNCERTAINTY

Inaccurate description of component characteristics, torn-and-worn effects on plant components, or shifting of operating points cause dynamic perturbations in many industrial control systems. Such perturbations may be represented by variations of certain system parameters over some possible value ranges. They affect the low-frequency range performance and are called "parametric uncertainties."

Studying an example is quite helpful. Assume the simple RLC circuit shown in Fig. 1.1.

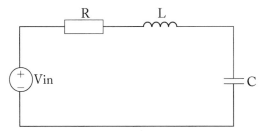

Figure 1.1: Typical RLC circuit.

The transfer function between the capacitor voltage and the input voltage can be written as:

$$\frac{v_c(s)}{v_{in}(s)} = \frac{\frac{1}{LC}}{s^2 + \frac{R}{L}s + \frac{1}{LC}} = \frac{a}{s^2 + bs + a}, \tag{1.1}$$

where $a = \frac{1}{LC}$ and $b = \frac{R}{L}$. R, L, and C can be written as $R = R_0 + \delta_R$, $L = L_0 + \delta_L$, and $C = C_0 + \delta_C$. δ_R, δ_L, and δ_C show the effect of aging, measurement error, replacement of the components, etc.

a and b can be written in the same way as $a_0 + \delta_a$ and $b_0 + \delta_b$, respectively. $a_0 = \frac{1}{L_0 C_0}$ and $b_0 = \frac{R_0}{L_0}$ show the nominal values of a and b, respectively. According to the values of R_0, L_0, C_0, δ_R, δ_L, and δ_C, a and b can be written as:

$$a_{\min} < a < a_{\max}$$
$$b_{\min} < b < b_{\max}. \tag{1.2}$$

So, Equation (1.1) no longer describes a single transfer function. It is a family of transfer functions with uncertain coefficients. The term, "Interval Plant Model" is used for such plants (i.e., a plant with uncertain coefficients). Lower/upper bound of coefficients variations is known.

1.2.2   UNSTRUCTURED UNCERTAINTY

Many dynamic perturbations that may occur in different parts of a system can, however, be lumped into one single perturbation block Δ, for instance, some unmodeled, high-frequency dynamics. This uncertainty representation is referred to as "unstructured" uncertainty. In the case of linear, time-invariant systems, the block Δ may be represented by an unknown transfer function matrix. The unstructured dynamics uncertainty in a control system can be described in different ways [Gu et al., 2013]. The most famous ones are additive and input/output multiplicative perturbation configurations. If $G_p(s)$, $G_o(s)$, and I show the perturbed system dynamics, a nominal model description of the physical system and identity matrix, respectively, then:

- additive perturbation: $G_p(s) = G_o(s) + \Delta(s)$,

- input multiplicative perturbation: $G_p(s) = G_o(s) \times [I + \Delta(s)]$, and

- output multiplicative perturbation: $G_p(s) = [I + \Delta(s)] \times G_o(s)$.

In Single Input Single Output (SISO) systems, there is no difference between input multiplicative perturbation and output multiplicative perturbation. In Multi Input Multi Output (MIMO) systems the two descriptions are not necessarily the same.

1.2.3   STRUCTURED UNCERTAINTY

In some problems the uncertain parts can be taken out from the dynamics and the whole system can be rearranged in a standard configuration of (upper) Linear Fractional Transformation $F(M, \Delta)$. The uncertian block Δ would then have the following general form:

$$\Delta = diag\left\{\delta_1 I_{r_1}, \delta_2 I_{r_2}, \delta_3 I_{r_3}, \ldots, \delta_s I_{r_s}, \Delta_1, \ldots, \Delta_f\right\}, \delta_i \in \mathbb{C}, \Delta_j \in \mathbb{C}^{m_j \times m_j},$$

where $\sum_{i=1}^{s} r_i + \sum_{j=1}^{f} m_j = n$ with n is the dimension of the block Δ. So, Δ consist of s repeated scalar blocks and f full blocks. The full blocks need not be square. Since the Δ considered has a certain structure, such description is called structured.

1.3   ROBUST CONTROL

Robust control is a design methodology that explicitly deals with uncertainty. Robust control designs a controller such that:

- some level of performance of the controlled system is guaranteed; and

- irrespective of the changes in the plant dynamics/process dynamics within a predefined class the stability is guaranteed.

Some of the well-known robust control design techniques are studied briefly below.

1.3.1   KHARITONOV'S THEOREM

Kharitonov's theorem is used to assess the stability of a dynamical system when the physical parameters of the system are uncertain. It can be considered as a generalization of Routh–Hurwitz stability test. Routh–Hurwitz is concerned with an ordinary polynomial, i.e., a polynomial with fixed coefficients, while Kharitonov's theorem can study the stability of polynomials with uncertain (varying) coefficients.

Kharitonov's theorem is an analysis tool more than a synthesis tool. Kharitonov's theorem can be used to tune simple controllers such as Proportional-Integral-Derivative (PID). Designing high-order controllers using Kharitonov's theorem is not so common.

Barmish [1993] and Bhattacharyya et al. [1995] are good references for control engineering applications of Kharitonov's thorem. Plenty of tools and related theorems are gathered there.

Kharitonov's theorem used to design robust controller for DC-DC converters. For instance, Bevrani et al. [2010] used Kharitonov's theorem to tune the PI controller of a quadratic buck converter. Chang [1995] designed a robust lead-lag controller for a buck converter.

1.3.2 H_∞ CONTROL

H_∞ techniques formulates the required design specifications (control goles) as an optimal control problem in the frequency domain. In order to do this, some fictitious weighting functions are added to the system model. Weighting functions are selected with respect to the required design specifications. Selection of weights is not a trivial task and requires some trial and error to obtain the desired specifications. In fact, the most crucial and difficult task in robust controller design is a choice of the weighting functions. Lundstron [1991], Skogesttad and Postlethwaite [2000], and Beaven [1996] give very general guidelines for selection of the weights. Donha and Katebi [2007] and Alfaro-Cid et al. [2008] used intelligent optimization methods (i.e., Genetic Algorithm) to find the best weighting functions.

The H_∞ design does not always ensure robust stability and robust performance of the closed-loop system. This is the main disadvantage of H_∞ design techniques.

H_∞ techniques are studied in many papers and books. Some of the most well known are introduced here. Zames [1981] is the pioneering work which introduced the H_∞ control. Kwakernak [1993] is a good tutorial paper on H_∞ control with some numeric examples. Zhou and Doyle [1997] and Green and Limbeer [2012] are general texts on robust control and studied the H_∞ control in detail. Gu et al. [2013] and Chiang et al. [2007] are good references to learn how to design H_∞ controllers using MATLAB®.

H_∞ techniques are applied to a number of DC-DC converters successfully. Naim et al. [1997] is the main antecedent in the use of H_∞ to DC-DC converters. It designed a H_∞ controller for a boost converter. Output impedance, audio susceptibility, phase margin, and bandwidth of the control loop are the usual measure of performance in the DC-DC switch mode Pulse Width Modulator (PWM) converters. Reduction of output impedance in non-minimum phase converters (such as boost and buck-boost) is achieved at the expense of phase-marging reduction. However, the H_∞ controller designed in Naim et al. [1997] minimizes the output impedance in a wide frequency range without decreasing the phase margin. This paper neglegted the system uncertainties. Khayat et al. [2017] studied the robust control of boost converters in presence of uncertainties. Shaw and Veerachary [2017] designed a H_∞ for a High Gain Boost Converter (HGBC). Vidal-Idiarte et al. [2003] designed a H_∞ controller to maximize the band width of the control loop with a perfect tracking of the desired output voltage for boost and buck-boost converters. Experimental results are compaerd with those obtained using Sliding Mode Control (SMC) and current peak control. Hernandez [2008] used the H_∞ loopshaping to design a controller for a buck-boost converter. Designed controller showed better performance in comparison with PID controller. Gadoura [2001] and Gadoura et al. [2002] designed H_∞ controller for paralleled buck converter operating in current-mode control (CMC) and voltage-mode control (VMC).

1.3.3 μ SYNTHESIS

The μ synthesis uses the D-K or μ-K iteration methods [Gu et al., 2013] to minimize the peak value of the structured singular value of the closed-loop transfer function matrix over the set of all stabilizing controllers K. The structured singular value of a closed-loop system transfer matrix $M(s)$, with uncertainty $\boldsymbol{\Delta}$ and singular values σ is defined as:

$$\|M\|_\mu = \mu_\Delta^{-1}(M) := \overset{\min}{\Delta \in \boldsymbol{\Delta}} \{\overline{\sigma}(\Delta) : \det(I - M\Delta) = 0\}.$$

Usually the controller designed using the μ synthesis has a high order which makes the implemetation difficult. A model order reduction procedure is usually required.

The H_∞ control design techniques consider the system uncertinty in the unstructured form so the controller designed using the H_∞ techniques is conservative. The μ synthesis considers the uncertainty structures so its output is less conservative [Bevrani et al., 1999].

μ synthesis is used to design controller for a number of DC-DC converters successfully. Buso [1996] modeled the parametric uncertainties of a buck-boost converter as unstructured and designed the controller using the μ synthesis.

Bevrani et al. [2004] designed a controller for a Quasi Resonant Converters (QRC) operating in Continuous Current Mode (CCM) using the μ synthesis. Robust control of a parallel resonant converter based on the μ synthesis is studied in Bu et al. [1997]. Wallis and Tymerski [2000] give the general guidelines for designing the robust controller for switch mode DC-DC converters using the μ synthesis.

Table 1.1 compares the different types of robust controllers [Bevrani et al., 1999].

Table 1.1: Comparison between linear robust controllers

Linear Robust Controller	Advantages	Disadvantages
Kharitonov's controller	• Simplicity of method • Lower degree of controller	• Direct reduction of output impedance is impossible
H_∞ controller	• Relatively low degree of controller • Direct reduction of output impedance is impossible	• Difficulty in determination of weighting functions • Synthesis conservatism
μ controller	• Non conservative • Direct reduction of output impedance is impossible • Stability robustness under a wider range of load variations	• Difficulty in determination of weighting functions • High degree of controller • Larger settling times

1.4   DYNAMICS OF A BUCK CONVERTER WITHOUT UNCERTAINTY

The schematic of a buck converter is shown in Fig. 1.2. The buck converter is composed of two switches: a MOSFET switch and a diode. In this schematic, Vg, rg, L, rL, C, rC, and R show input DC source, input DC source internal resistance, inductor, inductor Equivalent Series Resistance (ESR), capacitor, capacitor ESR, and load, respectively. iO is a fictitious current source added to the schematic in order to calculate the output impedance of converter. In this section we assume that converter works in Continuous Current Mode (CCM). MOSFET switch is controlled with the aid of a PWM controller. MOSFET switch keeps closed for $D.T$ seconds and $(1 - D).T$ seconds open. D and T show duty ratio and switching period, respectively.

When MOSFET is closed, the diode is opened (Fig. 1.3).

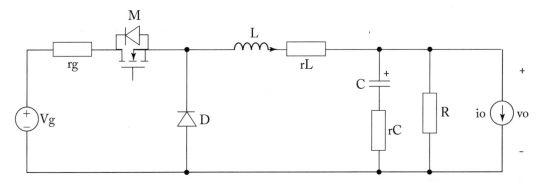

Figure 1.2: Schematic of a buck converter.

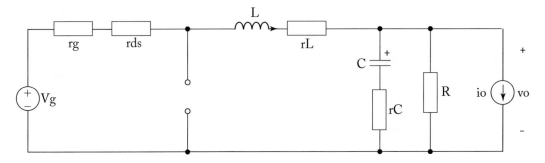

Figure 1.3: Equivalent circuit of a buck converter for closed MOSFET.

The circuit differential equations can be written as:

$$
\begin{cases}
\dfrac{di_L(t)}{dt} = \dfrac{1}{L}\left(-\left(r_g + r_{ds} + r_L + \dfrac{R \times r_C}{R + r_C}\right) i_L - \dfrac{R}{R + r_C} v_C + \dfrac{R \times r_C}{R + r_C} i_o + v_g\right) \\[4mm]
\dfrac{dv_C(t)}{dt} = \dfrac{1}{C}\left(\dfrac{R}{R + r_C} i_L - \dfrac{1}{R + r_C} v_C - \dfrac{R}{R + r_C} i_o\right)
\end{cases}
$$

$$
v_o = r_C C \dfrac{dv_C}{dt} + v_C = \dfrac{R \times r_C}{R + r_C} i_L + \dfrac{R}{R + r_C} v_C - \dfrac{R \times r_C}{R + r_C} i_o.
$$

When MOSFET is opened, the diode is closed (Fig. 1.4).

The circuit differential equations can be written as:

$$
\begin{cases}
\dfrac{di_L(t)}{dt} = \dfrac{1}{L}\left(-\left(r_D + r_L + \dfrac{R \times r_C}{R + r_C}\right) i_L - \dfrac{R}{R + r_C} v_C + \dfrac{R \times r_C}{R + r_C} i_o - v_D\right) \\[4mm]
\dfrac{dv_C(t)}{dt} = \dfrac{1}{C}\left(\dfrac{R}{R + r_C} i_L - \dfrac{1}{R + r_C} v_C - \dfrac{R}{R + r_C} i_o\right)
\end{cases}
$$

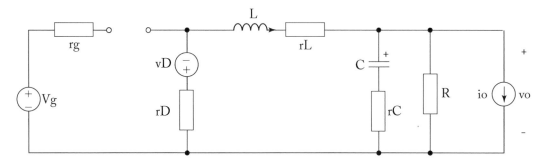

Figure 1.4: Equivalent circuit of a buck converter for opened MOSFET.

$$v_o = r_C C \frac{dv_C}{dt} + v_C = \frac{R \times r_C}{R + r_C} i_L + \frac{R}{R + r_C} v_C - \frac{R \times r_C}{R + r_C} i_O.$$

State Space Averaging (SSA) can be used to extract the small signal transfer functions of the DC-DC converter. The procedure of state space averaging is explained in detail in Suntio [2009] and Asadi and Eguchi [2018].

SSA has two important steps: averaging and linearizing the equivalent circuits dynamical equations. The SSA procedure can be summarized as follows.

Step 1 Circuit differential equations are written for different working modes (i.e., on/off state of semiconductor switches).

Step 2 Equations are time averaged over one period.

Step 3 Steady-state operating points are calculated by equating the derivative terms to zero.

Step 4 The averaged equations are linearized around the steady-state operating point found in Step 3.

Doing the SSA procedure manually is tedious and error prone. A program can be quite useful for this purpose. MATLAB® can be quite helpful for this purpose.

The following program shows the implemetation of SSA for a converter with component values, as shown in Table 1.2. The component values are assumed to be certain, i.e., have no uncertainty. Analysis results are shown in Figs. 1.5, 1.6, and 1.7.

```matlab
%This program extracts the small signal transfer functions
%of a Buck converter
clc
clear all

%converter components values
%fsw= 20 KHz
VG=50;      %input DC source voltage
rg=0.5;     %input DC source internal resistance
rds=0.04;   %MOSFET drain-source resistance
rD=0.01;    %Diode series resistance
VD=0.7;     %Diode voltage drop
rL=10e-3;   %Inductor Equivalent Series Resistance(ESR)
L=400e-6;   %Inductor value
rC=0.05;    %Capacitor ESR
C=100e-6;   %Capacitor value
R=20;       %Load resistor
D=0.4;      %Duty ratio
IO=0;       %Average value of output current source

syms iL vC io vg vD d
%iL : Inductor L1 current
%vC : Capacitor C1 voltage
%io : Output current source
%vg : Input DC source
%vD : Diode voltage drop
%d  : Duty cycle

%Closed MOSFET Equations
diL_dt_MOSFET_close=(-(rg+rds+rL+R*rC/(R+rC))*iL-R/(R+rC)*
   vC+R*rC/(R+rC)*io+vg)/L;
dvC_dt_MOSFET_close=(R/(R+rC)*iL-1/(R+rC)*vC-R/(R+rC)*io)/C;
vo_MOSFET_close=R*rC/(R+rC)*iL+R/(R+rC)*vC-R*rC/(R+rC)*io;

%Opened MOSFET Equations
diL_dt_MOSFET_open=(-(rD+rL+rC*R/(R+rC))*iL-R/(R+rC)*vC+R*
   rC/(R+rC)*io-vD)/L;
dvC_dt_MOSFET_open=(R/(R+rC)*iL-1/(R+rC)*vC-R/(R+rC)*io)/C;
```

```
vo_MOSFET_open=R*rC/(R+rC)*iL+R/(R+rC)*vC-R*rC/(R+rC)*io;

%Averaging
averaged_diL_dt=simplify(d*diL_dt_MOSFET_close+(1-d)*
    diL_dt_MOSFET_open);
averaged_dvC_dt=simplify(d*dvC_dt_MOSFET_close+(1-d)*
    dvC_dt_MOSFET_open);
averaged_vo=simplify(d*vo_MOSFET_close+(1-d)*
    vo_MOSFET_open);

%Substituting the steady values of: input DC voltage source,
%Diode voltage drop, Duty cycle and output current source
%and calculating the DC operating point(IL and VC)
right_side_of_averaged_diL_dt=subs(averaged_diL_dt,
    [vg vD d io],[VG VD D IO]);
right_side_of_averaged_dvC_dt=subs(averaged_dvC_dt,
    [vg vD d io],[VG VD D IO]);
DC_OPERATING_POINT=
solve(right_side_of_averaged_diL_dt==0,
    right_side_of_averaged_dvC_dt==0,'iL','vC');

IL=eval(DC_OPERATING_POINT.iL);
VC=eval(DC_OPERATING_POINT.vC);
VO=eval(subs(averaged_vo,[iL vC io],[IL VC IO]));

disp('Operating point of converter')
disp('----------------------------')
disp('IL(A)=')
disp(IL)
disp('VC(V)=')
disp(VC)
disp('VO(V)=')
disp(VO)
disp('----------------------------')

%Linearizing the averaged equations around the DC operating
%point. We want to obtain the matrix A,B,C and D.
%     x=Ax+Bu
%     y=Cx+Du
```

```
%
%where,
%    x=[iL vC]'
%    u=[io vg d]'
%since we used the variables D for steady state duty ratio
%and C to show the capacitors values we use AA, BB, CC
%and DD instead of A, B, C and D.

%Calculating the matrix A
A11=subs(simplify(diff(averaged_diL_dt,iL)),[iL vC d io],
    [IL VC D IO]);
A12=subs(simplify(diff(averaged_diL_dt,vC)),[iL vC d io],
    [IL VC D IO]);

A21=subs(simplify(diff(averaged_dvC_dt,iL)),[iL vC d io],
    [IL VC D IO]);
A22=subs(simplify(diff(averaged_dvC_dt,vC)),[iL vC d io],
    [IL VC D IO]);

AA=eval([A11 A12;
         A21 A22]);

%Calculating the matrix B
B11=subs(simplify(diff(averaged_diL_dt,io)),[iL vC d vD io vg],
    [IL VC D VD IO VG]);
B12=subs(simplify(diff(averaged_diL_dt,vg)),[iL vC d vD io vg],
    [IL VC D VD IO VG]);
B13=subs(simplify(diff(averaged_diL_dt,d)),[iL vC d vD io vg],
    [IL VC D VD IO VG]);

B21=subs(simplify(diff(averaged_dvC_dt,io)),[iL vC d vD io vg],
    [IL VC D VD IO VG]);
B22=subs(simplify(diff(averaged_dvC_dt,vg)),[iL vC d vD io vg],
    [IL VC D VD IO VG]);
B23=subs(simplify(diff(averaged_dvC_dt,d)),[iL vC d vD io vg],
    [IL VC D VD IO VG]);

BB=eval([B11 B12 B13;
         B21 B22 B23]);
```

```
%Calculating the matrix C
C11=subs(simplify(diff(averaged_vo,iL)),[iL vC d io],
    [IL VC D IO]);
C12=subs(simplify(diff(averaged_vo,vC)),[iL vC d io],
    [IL VC D IO]);

CC=eval([C11 C12]);

D11=subs(simplify(diff(averaged_vo,io)),[iL vC d vD io vg],
    [IL VC D VD IO VG]);
D12=subs(simplify(diff(averaged_vo,vg)),[iL vC d vD io vg],
    [IL VC D VD IO VG]);
D13=subs(simplify(diff(averaged_vo,d)),[iL vC d vD io vg],
    [IL VC VD IO VG]);

%Calculating the matrix D
DD=eval([D11 D12 D13]);

%Producing the State Space Model and obtaining the small
%signal transfer functions
sys=ss(AA,BB,CC,DD);
sys.inputname={'io';'vg';'d'};
sys.outputname={'vo'};

vo_io=tf(sys(1,1)); %Output impedance transfer function
                    %vo(s)/io(s)
vo_vg=tf(sys(1,2)); %vo(s)/vg(s)
vo_d=tf(sys(1,3));  %Control-to-output(vo(s)/d(s))

%drawing the Bode diagrams
figure(1)
bode(vo_io),grid minor,title('vo(s)/io(s)')

figure(2)
bode(vo_vg),grid minor,title('vo(s)/vg(s)')

figure(3)
bode(vo_d),grid minor,title('vo(s)/d(s)')
```

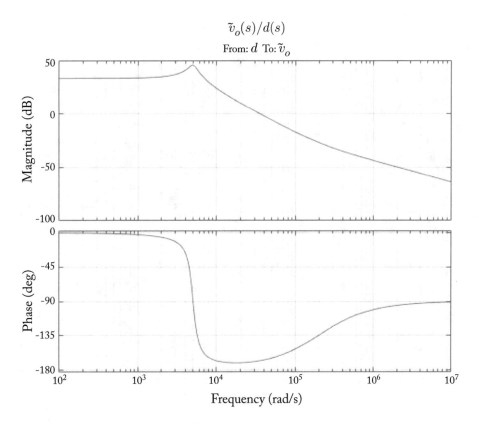

Figure 1.5: $\frac{\tilde{v}_o(s)}{\tilde{d}(s)} = 6257.7 \frac{s + 2 \times 10^5}{s^2 + 1203s + 2.523 \times 10^7}$ Bode diagram.

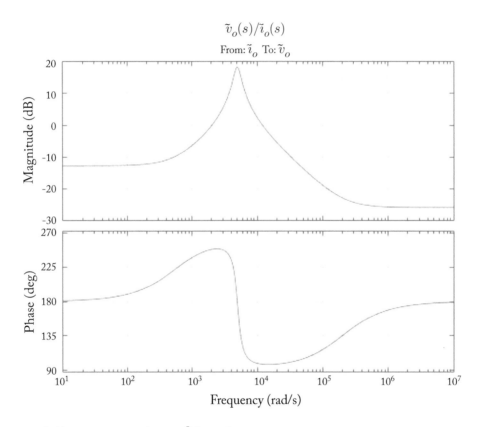

Figure 1.6: $\frac{\tilde{v}_o(s)}{\tilde{i}_o(s)} = -0.0499 \frac{(s+2\times10^5)(s+580)}{s^2+1203s+2.523\times10^7}$ Bode diagram.

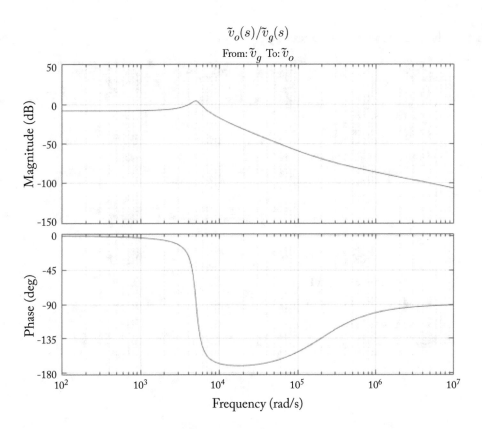

Figure 1.7: $\frac{\tilde{v}_o(s)}{\tilde{v}_g(s)} = 49.875\,\dfrac{(s+2\times10^5)}{s^2+1203s+2.523\times10^7}$ Bode diagram.

According to the analysis results, the converter with parameters given in Table 1.2 can be modeled, as shown in Fig. 1.8.

Table 1.2: The buck converter parameters (see Fig. 1.2)

	Nominal Value
Output voltage, Vo	20 V
Duty ratio, D	0.4
Input DC source voltage, Vg	50 V
Input DC source internal resistance, rg	0.5 Ω
MOSFET drain–source resistance, rds	40 mΩ
Capacitor, C	100 μF
Capacitor Equivaluent Series Resistance (ESR), rC	0.05 Ω
Inductor, L	400 μH
Inductor ESR, rL	10 mΩ
Diode voltage drop, vD	0.7 V
Diode forward resistance, rD	10 mΩ
Load resistor, R	20 Ω
Switching Frequency, Fsw	20 KHz

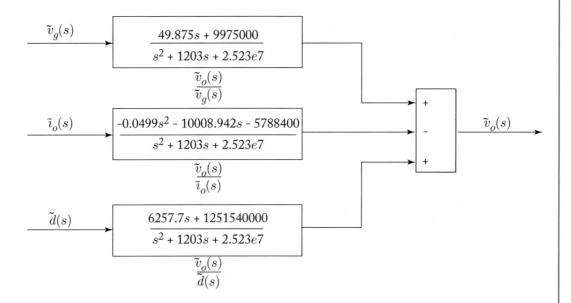

Figure 1.8: Dynamic model of the converter.

1.5   EFFECT OF COMPONENT VARIATIONS

Assume a buck converter composed of uncertain components. The component nominal values and their variations are given in Table 1.3.

Table 1.3: The buck converter parameters

	Nominal Value	Variations
Output voltage, Vo	20 V	0%
Input DC source voltage, Vg	50 V	±20%
Input DC source internal resistance, rg	0.5 Ω	±20%
MOSFET drain-source resistance, rds	40 mΩ	±20%
Capacitor, C	100 μF	±20%
Capacitor Equivaluent Series Resistance (ESR), rC	0.05 Ω	-10%, +90%
Inductor, L	400 μH	±10%
Inductor ESR, rL	10 mΩ	-10%, +90%
Diode voltage drop, vD	0.7 V	±30%
Diode forward resistance, rD	10 mΩ	-10%, +50%
Load resistor, R	20 Ω	±20%

We want to study the effect of these variations on the converter dynamics. Figures 1.9–1.11 show the changes in the transfer functions when parameters change according to Table 1.3.

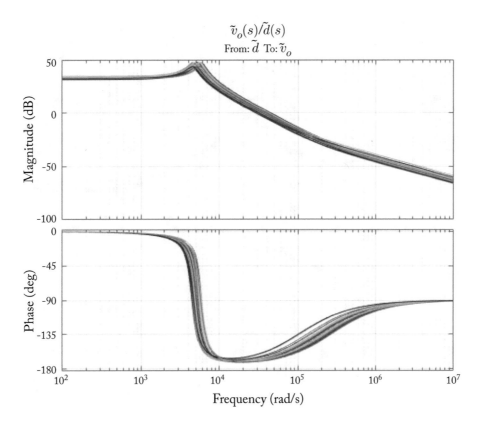

Figure 1.9: Effect of components changes on the $\frac{\tilde{v}_o(s)}{\tilde{d}(s)}$.

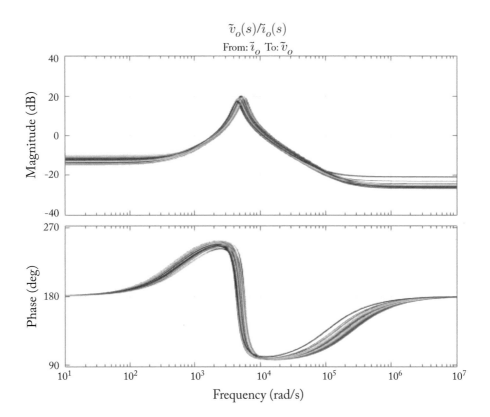

Figure 1.10: Effect of components changes on the $\frac{\tilde{v}_o(s)}{\tilde{i}_o(s)}$.

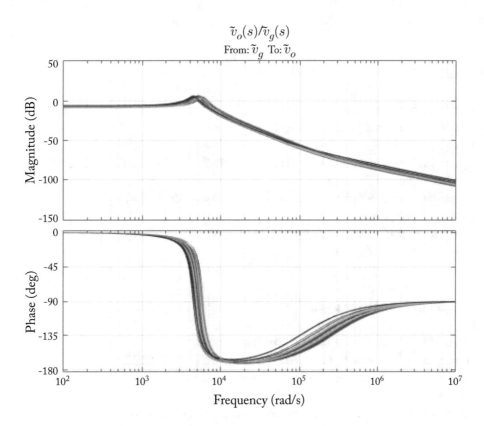

Figure 1.11: Effect of components changes on the $\frac{\tilde{v}_o(s)}{\tilde{v}_g(s)}$.

In order to design the robust controller using the Kharitonov's theorem, we need the lower/upper bound of transfer function coefficients. The following program extract the interval plant model of the converter with parameter variations, as shown in Table 1.3.

```
%This program calculates the small signal transfer functions
%of Buck converter and extracts the upper/lower bounds for
%transfer function coefficients. This program helps you design
%the controller using Kharitonov's theorem

clc
clear all

NumberOfIteration=150;
DesiredOutputVoltage=20;
s=tf('s');
vo_io_nominal=-0.049875*(s+2e5)*(s+580)/(s^2+1203*s+2.523e7);
    %Nominal vo(s)/io(s)
vo_vg_nominal=49.875*(s+2e5)/(s^2+1203*s+2.523e7);
    %Nominal vo(s)/vg(s)
vo_d_nominal=6257.7*(s+2e5)/(s^2+1203*s+2.523e7);
    %Nominal vo(s)/d(s)

n=0;
for i=1:NumberOfIteration
n=n+1;
%Definition of uncertainity in parameters
VG_unc=ureal('VG_unc',50,'Percentage',[-20 +20]);
    %Average value of input DC source is in the
    %range of 40..60
rg_unc=ureal('rds_unc',.5,'Percentage',[-20 +20]);
    %Input DC source resistance
rds_unc=ureal('rds_unc',.04,'Percentage',[-20 +20]);
    %MOSFET on resistance
C_unc=ureal('C_unc',100e-6,'Percentage',[-20 +20]);
    %Capacitor value
rC_unc=ureal('rC_unc',.05,'Percentage',[-10 +90]);
    %Capacitor Equivalent Series Resistance (ESR)
L_unc=ureal('L_unc',400e-6,'Percentage',[-10 +10]);
    %Inductor value
```

```matlab
rL_unc=ureal('rL_unc',0.01,'Percentage',[-10 +90]);
    %Inductor Equivalent Series Resistance (ESR)
rD_unc=ureal('rD_unc',.01,'Percentage',[-10 +50]);
    %Diode series resistance
VD_unc=ureal('VD_unc',.7,'Percentage',[-30 +30]);
    %Diode voltage drop
R_unc=ureal('R_unc',20,'Percentage',[-20 +20]);
    %Load resistance
IO=0;
    %Average value of output current source

%Sampling the uncertain set
%for instance usample(VG_unc,1) takes one sample of
%uncertain parameter VG_unc

VG=usample(VG_unc,1);
    %Sampled average value of input DC source
rg=usample(rg_unc,1);
    %Sampled internal resistance of input DC source
rds=usample(rds_unc,1);
    %Sampled MOSFET on resistance
C=usample(C_unc,1);
    %Sampled capacitor value
rC=usample(rC_unc,1);
    %Sampled capacitor Equivalent Series
Resistance(ESR)
L=usample(L_unc,1);
    %Sampled inductor value
rL=usample(rL_unc,1);
    %Sampled inductor Equivalent Series Resistance (ESR)
rD=usample(rD_unc,1);
    %Sampled diode series resistance
VD=usample(VD_unc,1);
    %Sampled diode voltage drop
R=usample(R_unc,1);
    %Sampled load resistance

%output voltage of an IDEAL (i.e., no losses) Buck
%converter operating in CCM is given by:
```

```
%VO=D.VG
%where
%VO: average value of output voltage
%D:   Duty Ratio
%VG: Input DC voltage
%So, for a IDEAL converter
%        VO
%D=----------
%        VG
%Since our converter has losses we use a bigger duty
%ratio, for instance:
%             VO
%D=1.05 --------
%             VG


D=1.05*DesiredOutputVoltage/(VG); % Duty cylcle

syms iL vC io vg vD d
%iL : Inductor L1 current
%vC : Capacitor C1 voltage
%io : Output current source
%vg : Input DC source
%vD : Diode voltage drop
%d  : Duty cycle

%Closed MOSFET Equations
diL_dt_MOSFET_close=(-(rg+rds+rL+R*rC/(R+rC))*iL-R/(R+rC)*
    vC+R*rC/(R+rC)*io+vg)/L;
dvC_dt_MOSFET_close=(R/(R+rC)*iL-1/(R+rC)*vC-R/(R+rC)*io)/C;
vo_MOSFET_close=R*rC/(R+rC)*iL+R/(R+rC)*vC-R*rC/(R+rC)*io;

%Opened MOSFET Equations
diL_dt_MOSFET_open=(-(rD+rL+rC*R/(R+rC))*iL-R/(R+rC)*vC+R*
    rC/(R+rC)*io-vD)/L;
dvC_dt_MOSFET_open=(R/(R+rC)*iL-1/(R+rC)*vC-R/(R+rC)*io)/C;
vo_MOSFET_open=R*rC/(R+rC)*iL+R/(R+rC)*vC-R*rC/(R+rC)*io;

%Averaging
averaged_diL_dt=simplify(d*diL_dt_MOSFET_close+(1-d)*
```

```
   diL_dt_MOSFET_open);
averaged_dvC_dt=simplify(d*dvC_dt_MOSFET_close+(1-d)*
   dvC_dt_MOSFET_open);
averaged_vo=simplify(d*vo_MOSFET_close+(1-d)*
   vo_MOSFET_open);

%Substituting the steady values of: input DC voltage
%source, Diode voltage drop, Duty cycle and output current
%source and calculating the DC operating point(IL and VC)
right_side_of_averaged_diL_dt=subs(averaged_diL_dt,
   [vg vD d io],[VG VD D IO]);
right_side_of_averaged_dvC_dt=subs(averaged_dvC_dt,
   [vg vD d io],[VG VD D IO]);

DC_OPERATING_POINT=
solve(right_side_of_averaged_diL_dt==0,
   right_side_of_averaged_dvC_dt==0,'iL','vC');

IL=eval(DC_OPERATING_POINT.iL);
VC=eval(DC_OPERATING_POINT.vC);
VO=eval(subs(averaged_vo,[iL vC io],[IL VC IO]));

disp('Operating point of converter')
disp('----------------------------')
disp('IL(A)=')
disp(IL)
disp('VC(V)=')
disp(VC)
disp('VO(V)=')
disp(VO)
disp('----------------------------')

%Linearizing the averaged equations around the DC
%operating point. We want to obtain the matrix A,B,C and D.
%      x=Ax+Bu
%      y=Cx+Du
%
%where,
%      x=[iL vC]'
```

```
%       u=[io vg d]'
%since we used the variables D for steady state duty
%ratio and C to show the capacitors values we use AA,
%BB, CC and DD instead of A, B, C and D.

%Calculating the matrix A
A11=subs(simplify(diff(averaged_diL_dt,iL)),[iL vC d io],
    [IL VC D IO]);
A12=subs(simplify(diff(averaged_diL_dt,vC)),[iL vC d io],
    [IL VC D IO]);

A21=subs(simplify(diff(averaged_dvC_dt,iL)),[iL vC d io],
    [IL VC D IO]);
A22=subs(simplify(diff(averaged_dvC_dt,vC)),[iL vC d io],
    [IL VC D IO]);

AA=eval([A11 A12;
         A21 A22]);

%Calculating the matrix B
B11=subs(simplify(diff(averaged_diL_dt,io)),
    [iL vC d vD io vg],[IL VC D VD IO VG]);
B12=subs(simplify(diff(averaged_diL_dt,vg)),
    [iL vC d vD io vg],[IL VC D VD IO VG]);
B13=subs(simplify(diff(averaged_diL_dt,d)),
    [iL vC d vD io vg],[IL VC D VD IO VG]);

B21=subs(simplify(diff(averaged_dvC_dt,io)),
    [iL vC d vD io vg],[IL VC D VD IO VG]);
B22=subs(simplify(diff(averaged_dvC_dt,vg)),
    [iL vC d vD io vg],[IL VC D VD IO VG]);
B23=subs(simplify(diff(averaged_dvC_dt,d)),
    [iL vC d vD io vg],[IL VC D VD IO VG]);

BB=eval([B11 B12 B13;
         B21 B22 B23]);

%Calculating the matrix C
C11=subs(simplify(diff(averaged_vo,iL)),[iL vC d io],
```

```
    [IL VC D IO]);
C12=subs(simplify(diff(averaged_vo,vC)),[iL vC d io],
    [IL VC D IO]);

CC=eval([C11 C12]);

D11=subs(simplify(diff(averaged_vo,io)),
    [iL vC d vD io vg],[IL VC D VD IO VG]);
D12=subs(simplify(diff(averaged_vo,vg)),
    [iL vC d vD io vg],[IL VC D VD IO VG]);
D13=subs(simplify(diff(averaged_vo,d)),
    [iL vC d vD io vg],[IL VC D VD IO VG]);

%Calculating the matrix D
DD=eval([D11 D12 D13]);

%Producing the State Space Model and obtaining the
%small signal transfer functions
sys=ss(AA,BB,CC,DD);
sys.inputname={'io';'vg';'d'};
sys.outputname={'vo'};

vo_io=tf(sys(1,1));
   %Output impedance transfer function vo(s)/io(s)
vo_vg=tf(sys(1,2));
   %vo(s)/vg(s)
vo_d=tf(sys(1,3));
   %Control-to-output (vo(s)/d(s))

%Extracts the transfer function coefficients
if n==1
      [num_vo_io,den_vo_io]=tfdata(vo_io,'v');
      [num_vo_vg,den_vo_vg]=tfdata(vo_vg,'v');
      [num_vo_d,den_vo_d]=tfdata(vo_d,'v');
else
      [num1,den1]=tfdata(vo_io,'v');
        %extracts the numerator and denominator
        %of vo(s)/io(s)
       num_vo_io=[num_vo_io;num1];
```

```
          %numerator of vo(s)/io(s)
        den_vo_io=[den_vo_io;den1];
         %denominator of vo(s)/io(s)

        [num2,den2]=tfdata(vo_vg,'v');
        num_vo_vg=[num_vo_vg;num2];
        den_vo_vg=[den_vo_vg;den2];

        [num3,den3]=tfdata(vo_d,'v');
        num_vo_d=[num_vo_d;num3];
        den_vo_d=[den_vo_d;den3];
end
disp('Percentage of work done:')
disp(n/NumberOfIteration*100)
   %shows the progress of the loop
disp('')
end
disp('')
disp('vo(s)/d(s)')
disp('maximum of numerator coefficients:')
disp(max(num_vo_d))
disp('minimum of numerator coefficients:')
disp(min(num_vo_d))
disp('')
disp('maximum of denominator coefficients:')
disp(max(den_vo_d))
disp('minimum of denominator coefficients:')
disp(min(den_vo_d))
disp('-------------')
disp('vo(s)/io(s)')
disp('maximum of numerator coefficients:')
disp(max(num_vo_io))
disp('minimum of numerator coefficients:')
disp(min(num_vo_io))
disp('')
disp('maximum of denominator coefficients:')
disp(max(den_vo_io))
disp('minimum of denominator coefficients:')
disp(min(den_vo_io))
```

```
disp('--------------')
disp('vo(s)/vg(s)')
disp('maximum of numerator coefficients:')
disp(max(num_vo_vg))
disp('minimum of numerator coefficients:')
disp(min(num_vo_vg))
disp('')
disp('maximum of denominator coefficients:')
disp(max(den_vo_vg))
disp('minimum of denominator coefficients:')
disp(min(den_vo_vg))
disp('-------------')
```

The program randomly samples the component values (according to the given variations) in each iteration. SSA is applied to the sampled values in order to obtain the transfer function coefficients. Obtained coefficients are stored in an array. The program calculates and shows the minimum/maximum value of each coefficient when iterations are finished.

The uncertain model of the studied Buck converter is shown in Table 1.4.

Table 1.4: Interval plant model of the transfer functions

$\dfrac{\tilde{v}_o(s)}{\tilde{d}(s)}$	$\dfrac{\tilde{v}_o(s)}{\tilde{i}_o(s)}$	$\dfrac{\tilde{v}_o(s)}{\tilde{v}_g(s)}$
$\dfrac{\tilde{v}_o(s)}{\tilde{d}(s)} = \dfrac{b_1 s + b_0}{s^2 + a_1 s + a_0}$	$\dfrac{\tilde{v}_o(s)}{\tilde{i}_o(s)} = \dfrac{b_2 s^2 + b_1 s + b_0}{s^2 + a_1 s + a_0}$	$\dfrac{\tilde{v}_o(s)}{\tilde{v}_g(s)} = \dfrac{b_1 s + b_0}{s^2 + a_1 s + a_0}$
$4.2793 \times 10^3 < b_1 < 1.1302 \times 10^4$	$-0.0883 < b_2 < -0.045$	$38.4197 < b_1 < 118.2969$
$7.9196 \times 10^8 < b_0 < 1.9482 \times 10^9$	$-1.2407 \times 10^4 < b_1 < -8.3366 \times 10^3$	$7.4277 \times 10^6 < b_0 < 1.6682 \times 10^7$
$1.0076 \times 10^3 < a_1 < 1.6746 \times 10^3$	$-1.0087 \times 10^7 < b_0 < -3.8357 \times 10^6$	$1.0076 \times 10^3 < a_1 < 1.6746 \times 10^3$
$1.9979 \times 10^7 < a_0 < 3.3958 \times 10^7$	$1.0076 \times 10^3 < a_1 < 1.6746 \times 10^3$	$1.9979 \times 10^7 < a_0 < 3.3958 \times 10^7$
	$1.9979 \times 10^7 < a_0 < 3.3958 \times 10^7$	

Step response of $\dfrac{\tilde{v}_o(s)}{\tilde{v}_g(s)}$ (with coefficients randomly selected within the allowed bounds) is shown in Fig. 1.12. This figure is produced with the aid of the following code.

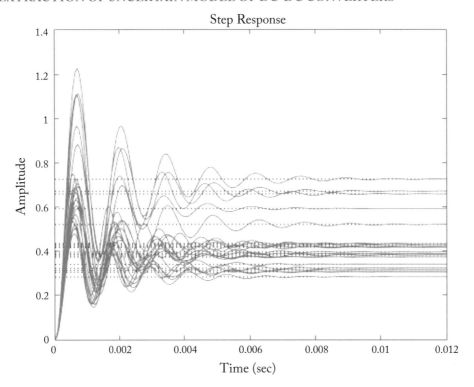

Figure 1.12: Step response of $\frac{\tilde{v}_o(s)}{\tilde{v}_g(s)}$ with randomly selected coefficients.

```
clc

%Defining the uncertain coefficients
b1_max=118.2969;
b1_nom=49.875;
b1_min=38.4197;
b1=ureal('b1',b1_nom,'Range',[b1_min, b1_max]);

b0_max=1.6682e7;
b0_nom=9975000;
b0_min=7.4277e6;
b0=ureal('b0',b0_nom,'Range',[b0_min, b0_max]);

a1_max=1.6746e3;
a1_nom=1203;
```

```
a1_min=1.0076e3;
a1=ureal('a1',a1_nom,'Range',[a1_min, a1_max]);

a0_max=3.3958e7;
a0_nom=2.523e7;
a0_min=1.9979e7;
a0=ureal('a0',a0_nom,'Range',[a0_min, a0_max]);

%uncertain model of vo(s)/vg(s)
H=tf([b1 b0],[1 a1 a0]);

%Step response of H
step(H)
```

1.6 CONCLUSION

This chapter showed how one can obtain the interval plant model of a DC-DC converter. A buck converter is studied as an illustrative example. The proposed method can be used to extract the uncertain model of other types of DC-DC converters. The next chapter introduces Kharitonov's theorem.

REFERENCES

Alfaro-Cid, E., McGookin, E., and Murray-Smith, D. Optimisation of the weighting functions of an H_∞ controller using genetic algorithms and structured genetic algorithms. *International Journal of System and Science*, pp. 335–347, 2008. DOI: 10.1080/00207720701777959. 4

Asadi, F. and Eguchi, K. *Dynamics and Control of DC-DC Converters*, Morgan & Claypool, 2018. DOI: 10.2200/s00828ed1v01y201802pel010. 8

Barmish, R. *New Tools for Robustness of Linear Systems*, Macmillan, 1993. 4

Beaven, R., Wright, M., and Seaward, D. Weighting function selection in the H_∞ design process. *Control Engineering Practice*, pp. 625–633, 1996. DOI: 10.1016/0967-0661(96)00044-5. 4

Bevrani, H., Abrishamchian, M., and Safari-Shad, N. Nonlinear and linear robust control of switching power converters. *Proc. of the IEEE International Conference on Control Applications*, pp. 808–813, 1999. DOI: 10.1109/cca.1999.807765. 5, 6

Bevrani, H., Babahajyani, P., Habibi, F., and Hiyama, T. Robust control design and implementation for a quadratic buck converter. *The International Power Electronics Conference ECCE ASIA*, pp. 99–103, Sapporo, 2010. DOI: 10.1109/ipec.2010.5543644. 4

Bevrani, H., Ise, T., Mitani, Y., et al. A robust approach to controller design for DC-DC quasi-resonant converter. *IEEJ Transaction on Industry Application*, pp. 91–100, 2004. DOI: 10.1541/ieejias.124.91. 6

Bhattacharyya, S., Chapellat, H., and Keel, L. *Robust Control the Parametric Approach*, Prentice Hall PTR, 1995. DOI: 10.1016/b978-0-08-042230-5.50016-5. 4

Bu, J., Sznaier, M., Wang, Z., et al. Robust control design for a parallel resonant converter using μ synthesis. *IEEE Transaction on Power Electronics*, pp. 837–853, 1997. DOI: 10.1109/63.623002. 6

Buso, S. Synthesis of a robust voltage controller for a buck-boost converter. *Proc. IEEE Power Electronics Specialists Conference (PESC)*, pp. 766–772, 1996. DOI: 10.1109/pesc.1996.548668. 6

Chang, C. Robust control of DC-DC converters: The buck converter. *Power Electronics Specialists Conference*, pp. 1094–1097, Atlanta, 1995. DOI: 10.1109/pesc.1995.474951. 4

Chiang, R., Safonov, M., Balas, G., and Packard, A. *Robust Control Toolbox*, 3rd ed., The Mathworks, Inc., 2007. 5

Donha, D. and Katebi, M. Automatic weight selection for H_∞ controller synthesis. *International Journal of Systems Science*, pp. 651–664, 2007. DOI: 10.1080/00207720701500559. 4

Erikson, R. and Maksimovic, D. *Fundametals of Power Electronics*, Springer, 2007. DOI: 10.1007/b100747. 1

Gadoura, I., Suntio, T., and Zenger, K. Dynamic system modeling and analysis for multiloop operation of paralleled DC/DC converters. *Proc. of the International Conference on Power Electronics and Intelligent Motion*, pp. 443–448, 2001. 5

Gadoura, I., Suntio, T., and Zenger, K. Model uncertainty and robust control of paralleled DC/DC converters. *International Conference on Power Electronics, Machines and Drives*, pp. 74–79, 2002. DOI: 10.1049/cp:20020092. 5

Green, M. and Limbeer, D. *Linear Robust Control*, Dover Publications, 2012. 5

Gu, D., Petkov, P., and Konstantinov, M. *Robust Control Design with MATLAB*, Springer, 2013. DOI: 10.1007/978-1-4471-4682-7. 3, 5

Hernandez, W. Robust control applied to improve the performance of a buck-boost converter. *WSEAS Transaction on Circuit and Systems*, pp. 450–459, 2008. 5

Khayat, Y., Naderi, M., Shafiee, Q., et al. Robust control of a DC-DC boost converter: H_2 and H_∞ techniques. *8th Power Electronics, Drive Systems and Technologies Conference (PEDSTC)*, pp. 407–412, 2017. DOI: 10.1109/pedstc.2017.7910360. 5

Kislovski, A. S., Ridl, R., and Socal, N. *Dynamic Analysis of Switching Mode DC-DC Converter*, Van Nostrand Reinhold, New York, 1991. DOI: 10.1007/978-94-011-7849-5. 1

Kwakernak, H. Robust control and H_∞ optimization. *Automatica*, pp. 255–273, 1993. 5

Lundstron, P., Skogestad, S., and Wang, Z. Performance weight selection for H-infinity and μ-control methods. *Transactions of the Institute of Measurement and Control*, (24), pp. 1–252, 1991. DOI: 10.1177/014233129101300504. 4

Maksimovic, D., Stankovic, A., Tottuvelil, V., et al. Modeling and simulation of power electronic converters. *Proc. IEEE*, pp. 898–912, 2001. DOI: 10.1109/5.931486. 1

Middlebrook, R. and Cuk, S. General unified approach to modelling switching-converter power stages. *International Journal of Electronics Theoretical and Experimental*, 42(6):521–550, 1977. DOI: 10.1109/pesc.1976.7072895. 1

Naim, R., Weiss, G., and Ben-Yaakov, S. H_∞ control applied to boost power converters. *IEEE Transactions on Power Electron*, pp. 677–683, 1997. DOI: 10.1109/63.602563. 5

Shaw, P. and Veerachary, M. Mixed-sensitivity based robust H_∞ controller design for high-gain boost converter. *International Conference on Computer, Communications and Electronics (Comptelix)*, pp. 612–617, 2017. DOI: 10.1109/comptelix.2017.8004042. 5

Skogesttad, S. and Postlethwaite, I. *Multivariable Feedback Control-Analysis and Design*, John Wiley & Sons, 2000. 4

Suntio, T. *Dynamic Profile of Switched Mode Converters: Modeling, Analysis and Control*, Wiley WCH, 2009. DOI: 10.1002/9783527626014. 8

Vidal-Idiarte, E., Martinez-Salamero, L., Valderrama-Blavi, H., Guinjoan, F., and Maixe, J. Analysis and design of H_∞ control of nonminimum phase-switching converters. *IEEE Transactions on Circuits and Systems I: Fundamental Theory and Applications*, pp. 1316–1323, 2003. DOI: 10.1109/tcsi.2003.816337. 5

Wallis, G. and Tymerski, R. Generalized approach for μ-synthesis of robust switching regulators. *IEEE Transactions on Aerospace Electronic Systems*, pp. 422–431, 2000. DOI: 10.1109/7.845219. 6

Zames, G. Feedback and optimal sensitivity: Model reference transformations, multiplicative seminorms, and approximate inverses. *IEEE Transactions on Automatic Control*, pp. 301–320, 1981. DOI: 10.1109/tac.1981.1102603. 5

Zhou, K. and Doyle, J. *Essential of Robust Control*, Pearson, 1997. 5

CHAPTER 2

Overview of Kharitonov's Theorem

2.1 INTRODUCTION

Kharitonov's theorem addresses the stability problem of interval polynomials. It can be considered as a generalization of Routh–Hurwitz stability test. Routh–Hurwitz is concerned with an ordinary polynomial, i.e., a polynomial with fixed coefficients, while Kharitonov's theorem can study the stability of polynomials with uncertain coefficients. Before presenting the theorem, some definition must be introduced.

2.2 KHARITONOV'S THEOREM AND RELATED MATHEMATICAL TOOLS

Before presenting the Kharitonov's theorem, some definitions must be studied. Required definitions are given below.

Definition 2.1 (*Stability*). A fixed polynomial $P(s)$, i.e., with fixed coefficients, is said to be stable (or Hurwitz) if all its roots lie in the strict Left Half Plane (LHP). □

For example, $p(s) = s^2 + 2s + 5$ is stable since its roots $(-1 \pm 2j)$, lie in the LHP. Dependence of a transfer function on a vector of uncertain parameters q is shown with the aid of $p(s, q)$ instead of $p(s)$.

Definition 2.2 (*Robust Stability*). A given family of polynomials $P = \{p(\cdot, q) : q \in Q\}$ is said to be robustly stable if, for all $q \in Q$, $P(s, q)$ is stable. That is, for all roots of $P(s, q)$ lie in the strict LHP. □

Definition 2.3 (*D Stability*). A polynomial is said to be D stable if all of its roots lie in $D \subseteq C$. □

In the above definition, D is sub-region of LHP. In fact, D stability is an attempt to restrict the acceptable region of poles. Assume a system that needs a settling time of less than 5 sec and

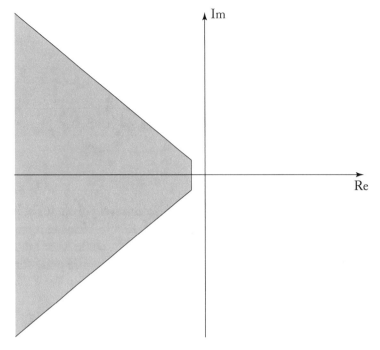

Figure 2.1: Acceptable region for close loop poles.

damping greater than 0.707. The acceptable region of close loop poles for such a system is no longer LHP. The acceptable region is shown in Fig. 2.1.

Definition 2.4 (*Interval Polynomial*).   An interval polynomial is the family of all polynomials

$$p(s) = a_0 + a_1 s^1 + a_2 s^2 + \cdots + a_n s^n, \tag{2.1}$$

where $\forall i, a_i \in [l_i, u_i]$ and $0 \notin [l_n, u_n]$. $\square$

In the above definition, l_i and u_i denote the bounding interval (i.e., minimum and maximum) for the i-th coefficient. $0 \notin [l_n, u_n]$ keeps the degree of polynomial family constant, i.e., all the members are polynomial of degree n.

For example, $p(s, q) = [1, 2] s^5 + [3, 4] s^4 + [5, 6] s^3 + [7, 8] s^2 + [9, 10] s + [11, 12]$ is an interval polynomial; it contains an infinite number of polynomials, for instance, $p(s, q) = 1.2 s^5 + 3.9 s^4 + 5 s^3 + 7.1 s^2 + 9.13 s + 12$ is a member of this interval polynomial.

Definition 2.5 (*Invariant Degree*).   A family of polynomials given by $P = \{p(\cdot, q) : q \in Q\}$ is said to have invariant degree if: given any $q^1, q^2 \in Q$, it follows that:

$$\deg p(s, q^1) = \deg p(s, q^2), \tag{2.2}$$

where deg shows the degree of polynomial, i.e., maximum power available in the polynomial. □

Definition 2.6 (*Kharitonov's Polynomials*). Associated with the interval polynomial $p(s, q) = \sum_{i=0}^{n} [q_i^-, q_i^+] s^i$ are the for fixed polynomials:

$$K_1(s) = q_0^- + q_1^- s + q_2^+ s^2 + q_3^+ s^3 + q_4^- s^4 + q_5^- s^5 + q_6^+ s^6 + \dots$$
$$K_2(s) = q_0^+ + q_1^+ s + q_2^- s^2 + q_3^- s^3 + q_4^+ s^4 + q_5^+ s^5 + q_6^- s^6 + \dots$$
$$K_3(s) = q_0^+ + q_1^- s + q_2^- s^2 + q_3^+ s^3 + q_4^+ s^4 + q_5^- s^5 + q_6^- s^6 + \dots$$
$$K_4(s) = q_0^- + q_1^+ s + q_2^+ s^2 + q_3^- s^3 + q_4^- s^4 + q_5^+ s^5 + q_6^+ s^6 + \dots$$

(2.3)

□

For example, if

$$p(s, q) = [1, 2]s^5 + [3, 4]s^4 + [5, 6]s^3 + [7, 8]s^2 + [9, 10]s + [11, 12].$$

Then:

$$K_1(s) = 11 + 9s + 8s^2 + 6s^3 + 3s^4 + s^5,$$
$$K_2(s) = 12 + 10s + 7s^2 + 5s^3 + 4s^4 + 2s^5,$$
$$K_3(s) = 12 + 9s + 7s^2 + 6s^3 + 4s^4 + s^5,$$
$$K_5(s) = 11 + 10s + 8s^2 + 5s^3 + 3s^4 + 2s^5.$$

(2.4)

Theorem 2.7 (*Kharitonov's Theorem (1978a)*). *An interval polynomial family P with invariant degree is robustly stable if and only if its four Kharitonov's polynomials are stable.* □

What is somewhat surprising about Kharitonov's theorem is that the stability problem of an infinite number of polynomials is reduced to stability problem of four fixed polynomials. The stability of these four polynomials can be tested via Routh–Hurwitz or any other method, i.e., using a software package.

For example, if

$$p(s, q) = [1, 2]s^5 + [3, 4]s^4 + [5, 6]s^3 + [7, 8]s^2 + [9, 10]s + [11, 12].$$

(2.5)

Then,

$$K_1(s) = 11 + 9s + 8s^2 + 6s^3 + 3s^4 + s^5,$$
$$K_2(s) = 12 + 10s + 7s^2 + 5s^3 + 4s^4 + 2s^5,$$
$$K_3(s) = 12 + 9s + 7s^2 + 6s^3 + 4s^4 + s^5,$$
$$K_5(s) = 11 + 10s + 8s^2 + 5s^3 + 3s^4 + 2s^5.$$

(2.6)

Using the classical Routh–Hurwitz table, it is easy to verify that all four polynomial have zero in Right Half Plane (RHP). So, using the Kharitonov's theorem, all the members of the afore-mentioned interval polynomial are not stable.

Kharitonov's theorem can be extended to polynomial with complex coefficients. Assume q_i and r_i denote the uncertainty in the real and imaginary parts of the coefficients of s^i, respectively, i.e.,

$$p(s,q,r) = \sum_{i=0}^{n}(q_i + jr_i)s^i. \tag{2.7}$$

Assume Q and R as uncertainty bounding sets for q and r, respectively. $P = \{p(\cdot,q,r) : q \in Q, r \in R\}$ is a complex coefficient interval polynomial family. Like real coefficient case $q_i^- \le q_i \le q_i^+, r_i^- \le r_i \le r_i^+$

$$p(s,q,r) = \sum_{i=0}^{n}\left(\left[q_i^-,q_i^+\right] + j\left[r_i^-,r_i^+\right]\right)s^i. \tag{2.8}$$

Lemma 2.8 *The nth order interval polynomial $p(s,q) = \sum_{i=0}^{n}\left[q_i^-,q_i^+\right]s^i$ is robustly stable if and only if the following is true.*

- *For $n = 3$, $K_3(s)$ is stable.*

- *For $n = 4$, $K_2(s)$ and $K_3(s)$ are stable.*

- *For $n = 5$, $K_2(s)$, $K_3(s)$ and $K_4(s)$ are stable.*

- *For $n \ge 6$, $(K_1(s)$, $K_2(s)$, $K_3(s)$ and $K_4(s)$ are stable.*

For example, testing the stability of $p(s,q) = [1,1]s^3 + [2,3]s^2 + [0.5,1]s + [1,2]$ can be done by testing the stability of $K_3(s) = s^3 + 2s^2 + 0.5s + 2$. Since $K_3(s) = s^3 + 2s^2 + 0.5s + 2$ is stable, the given family is stable.

Definition 2.9 (*Complex Coefficient Kharitonov's Polynomials*).   Associated with complex coefficient interval polynomials given as

$$p(s,q,r) = \sum_{i=0}^{n}\left(\left[q_i^-,q_i^+\right] + j\left[r_i^-,r_i^+\right]\right)s^i \tag{2.9}$$

are eight fixed Kharitonov's polynomials:

$$K_1{}^+(s) = \left(q_0{}^- + jr_0{}^-\right) + \left(q_1{}^- + jr_1{}^+\right)s + \left(q_2{}^+ + jr_2{}^+\right)s^2 + \left(q_3{}^+ + jr_3{}^-\right)s^3 + \dots$$
$$K_2{}^+(s) = \left(q_0{}^+ + jr_0{}^+\right) + \left(q_1{}^+ + jr_1{}^-\right)s + \left(q_2{}^- + jr_2{}^-\right)s^2 + \left(q_3{}^- + jr_3{}^+\right)s^3 + \dots$$
$$K_3{}^+(s) = \left(q_0{}^+ + jr_0{}^-\right) + \left(q_1{}^- + jr_1{}^-\right)s + \left(q_2{}^+ + jr_2{}^+\right)s^2 + \left(q_3{}^+ + jr_3{}^+\right)s^3 + \dots$$
$$K_4{}^+(s) = \left(q_0{}^- + jr_0{}^+\right) + \left(q_1{}^+ + jr_1{}^+\right)s + \left(q_2{}^+ + jr_2{}^-\right)s^2 + \left(q_3{}^- + jr_3{}^-\right)s^3 + \dots$$
$$K_1{}^-(s) = \left(q_0{}^- + jr_0{}^-\right) + \left(q_1{}^+ + jr_1{}^-\right)s + \left(q_2{}^+ + jr_2{}^+\right)s^2 + \left(q_3{}^- + jr_3{}^+\right)s^3 + \dots$$
$$K_2{}^-(s) = \left(q_0{}^+ + jr_0{}^+\right) + \left(q_1{}^- + jr_1{}^+\right)s + \left(q_2{}^- + jr_2{}^-\right)s^2 + \left(q_3{}^+ + jr_3{}^-\right)s^3 + \dots$$
$$K_3{}^-(s) = \left(q_0{}^+ + jr_0{}^-\right) + \left(q_1{}^+ + jr_1{}^+\right)s + \left(q_2{}^- + jr_2{}^+\right)s^2 + \left(q_3{}^- + jr_3{}^-\right)s^3 + \dots$$
$$K_4{}^-(s) = \left(q_0{}^- + jr_0{}^+\right) + \left(q_1{}^- + jr_1{}^-\right)s + \left(q_2{}^+ + jr_2{}^-\right)s^2 + \left(q_3{}^+ + jr_3{}^+\right)s^3 + \dots$$

$$(2.10)$$

☐

Theorem 2.10   (*Kharitonov's Theorem (1978b)*).   *A complex coefficient interval polynomial family* $P = \{p(\cdot, q, r) : q \in Q, p \in P\}$ *with invariant degree is robustly stable if its eight Kharitonov's polynomials are stable.* ☐

2.2.1   KHARITONOV'S RECTANGLE

Given an interval polynomial $p(s, q) = \sum_{i=0}^{n}[q_i{}^-, q_i{}^+]$ and a fixed frequency $\omega = \omega_0$, the set of possible values of $p(j\omega_0, q)$ is a rectangle called Kharitonov rectangle. The vertices of $p(j\omega_0, q)$ are given by four fixed Kharitonov's polynomials $K_1(s)$, $K_2(s)$, $K_3(s)$, and $K_4(s)$ at $s = j\omega_0$. Figure 2.2 shows the Kharitonov's rectangle.

For example, Kharitonov's rectangles of $p(s, q) = [0.25, 1.25]s^3 + [2.75, 3.25]s^2 + [0.75, 1.25]s + [0.25, 1.25]$ are drawn with the aid of the following MATLAB® code. The result is shown in Fig. 2.3.

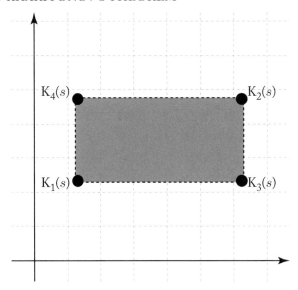

Figure 2.2: Kharitonov rectangle associated with $p(j\omega_0, q)$.

```
clc
syms s
K1=.25+.75*s+3.25*s^2+1.25*s^3;
K2=1.25+1.25*s+2.75*s^2+.25*s^3;
K3=1.25+.75*s+2.75*s^2+1.25*s^3;
K4=.25+1.25*s+3.25*s^2+.25*s^3;
j=sqrt(-1);

for w=0:.01:10

K1_R=eval(real(subs(K1,s,j*w)))
K1_I=eval(imag(subs(K1,s,j*w)))

K2_R=eval(real(subs(K2,s,j*w)))
K2_I=eval(imag(subs(K2,s,j*w)))

K3_R=eval(real(subs(K3,s,j*w)))
K3_I=eval(imag(subs(K3,s,j*w)))

K4_R=eval(real(subs(K4,s,j*w)))
```

```
K4_I=eval(imag(subs(K4,s,j*w)))

K_R=[K1_R K3_R K2_R K4_R];
K_I=[K1_I K3_I K2_I K4_I];

axis([-3 1.5 -0.6 1])
rectangle('Position',[K1_R,K1_I,K3_R-K1_R,K4_I-K1_I])
hold on
end
hold on
grid on
plot(0,0,'r*')
```

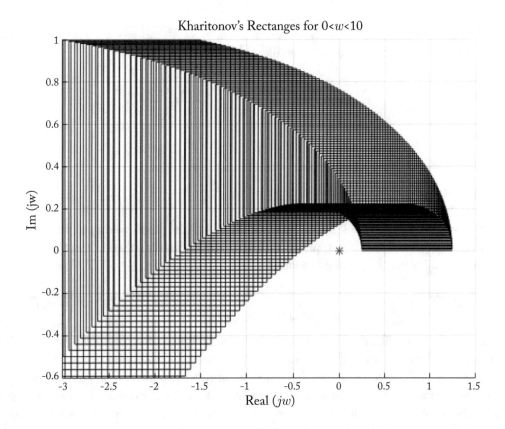

Figure 2.3: Kharitonov's rectangle for $p(s,q) = [0.25, 1.25]s^3 + [2.75, 3.25]s^2 + [0.75, 1.25]s + [0.25, 1.25]$.

Origin is shown with a red star in Fig. 2.3. Kharitonov's rectangle does not contain the origin, so the given polynomial is stable.

Lemma 2.11 (**Zero Exclusion Condition**)   *Suppose that $P = \{p\,(\cdot, q) : q \in Q\}$ is an interval polynomial with invariant degree and has at least one stable member. P is robustly stable if origin of complex plane, i.e., $Z = (0, 0)$, is excluded from the Kharitonov's rectangles for all $\omega \geq 0$.* □

2.3   CONTROLLER DESIGN FOR INTERVAL PLANTS

Assume a transfer function that numerator and denominator coefficients are uncertain:

$$p\,(s, q, r) = \frac{N(s, q)}{D(s, r)} = \frac{\sum_{i=0}^{m} \left[q_i{}^-, q_i{}^+\right] s^i}{s^n + \sum_{i=0}^{n-1} \left[r_i{}^-, r_i{}^+\right] s^i}. \tag{2.11}$$

Assume that $p(s, q, r)$ is strictly proper, that is $m < n$. We call $p\,(s, q, r)$ interval *plant* for obvious reasons. Assume a control scheme like that shown in Fig. 2.4.

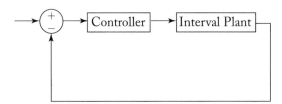

Figure 2.4: Feedback control of interval plant $p\,(s, q, r)$.

We want to find a controller $C(s)$ which stabilizes all the plants belong to $p\,(s, q, r)$. This problem can be solved using two approaches: applying Kharitonov's theorem to the close-loop denominator and using the 16-plant theorem.

2.3.1   FIRST APPROACH: APPLYING KHARITONOV'S THEOREM TO THE CLOSED LOOP DENOMINATOR

Assume a control loop like that shown in Fig. 2.4. The designer must choose a controller. Generally, a first-order controllers like PI, lead, or lag is selected. Close-loop transfer function (in presence of selected controller) is calculated next. Controller parameters (i.e., proportional gain and integral gain for PI controllers) are selected such that the close-loop denominator is robustly stable with respect to Kharitonov's theorem. Selection of suitable control parameters is done with the aid of loop commands in software environments.

For example, if interval plant is given by $H_P\,(s) = \frac{as+b}{s^2+cs+d}$ with $1 < a < 2$, $3 < b < 4$, $-5 < c < 6$, and $7 < d < 8$ and the designer chooses a PI controller with general form of

$H_C(s) = K_p + \frac{K_I}{S}$, then the close-loop transfer function is:

$$H_{CL}(s) = \frac{H_P(s) \times H_C(s)}{1 + H_P(s) \times H_C(s)} = \frac{\frac{as+b}{s^2+cs+d} \times \left(K_p + \frac{K_I}{s}\right)}{1 + \frac{as+b}{s^2+cs+d} \times \left(K_p + \frac{K_I}{s}\right)}$$

$$= \frac{(as+b)(K_I + K_P s)}{s^3 + (c + aK_p)s^2 + (d + aK_I + bK_P)s + bK_I}. \qquad (2.12)$$

The denominator of obtained transfer function is $s^3 + (c + aK_p)s^2 + (d + aK_I + bK_P)s + bK_I$. K_P and K_I are design parameters and must be find. Table 2.1 shows the minimum and maximum values of denominator coefficients with respect to uncertainty bounds given for $a, b, c,$ and d.

Table 2.1: Minimum and maximum values of denominator coefficients

Term	Minimum	Maximum
$1s^3$	1	1
$(c + aK_p)s^2$	$-5 + 1 \times K_p$	$6 + 2 \times K_p$
$(d + aK_I + bK_p)s^1$	$7 + 1 \times K_I + 3 \times K_p$	$8 + 2 \times K_I + 4 \times K_p$
$bK_I s^0$	$3 \times K_I$	$4 \times K_I$

Based on the minimum and maximum values of coefficients shown in Table 2.1, four Kharitonov's polynomials are formed:

$$\begin{aligned}
K_1(s) &= 3K_I + (7 + K_I + 3K_P)s + (6 + 2K_P)s^2 + s^3, \\
K_2(s) &= 4K_I + (8 + 2K_I + 4K_P)s + (-5 + 1K_P)s^2 + s^3, \\
K_3(s) &= 4K_I + (7 + 1K_I + 3K_P)s + (6 + 2K_P)s^2 + s^3, \\
K_4(s) &= 3K_I + (8 + 2K_I + 4K_P)s + (-5 + 1K_P)s^2 + s^3.
\end{aligned} \qquad (2.13)$$

Obtaining the suitable values for K_P and K_I is done with the aid of nested loop. First, a reasonable range and step for K_P and K_I are determined. Using nested loops suitable values of K_P and K_I are found.

Since in this example the denominator $(s^3 + (c + aK_p)s^2 + (d + aK_I + bK_P)s + bK_I)$ is a third-order polynomial, it is enough to stabilize $K_3(s)$ (see Lemma 2.8). The following pseudo-code shows this step:

for $K_P = K_{P,min} : K_{P,step} : K_{P,max}$
    for $K_I = K_{I,min} : K_{I,step} : K_{I,max}$

if $K_3(s)$ is Hurwitz, put the K_P and K_I in the acceptable set.

next K_I

next K_P

When the number of controller parameters increase, the number of required nested loops increase as well. In each iteration, polynomial roots must be calculated which is a time-consuming task. That's why a designer generally takes the simple forms like P, PI, lead, and lag for the controller to keep the required processing time reasonable.

Obtained set (output of nested loops) is the set of parameters that robustly stabilize the close loop system of Fig. 2.4.

2.3.2 SECOND APPROACH: 16-PLANT THEOREM

Assume a first-order compensator $C(s) = \frac{N_C(s)}{D_C(s)} = k\frac{s-z}{s-p}$. So, a close-loop polynomial is $p(s, q, r) = k(s-z)N(s, q) + (s-p)D(s, r)$. This leads to the family of close-loop polynomials:

$$P_{CL} = \{p(\cdot, q, r) : q \in Q, r \in R\}. \tag{2.14}$$

Definition 2.12 (*16 Kharitonov's Plants*).   Given an interval plant family P with Kharitonov's polynomials $N_1(s)$, $N_2(s)$, $N_3(s)$, $N_4(s)$, and $D_1(s)$, $D_2(s)$, $D_3(s)$, and $D_4(s)$ for the numerator and denominator, respectively. Sixteen Kharitonov's plants are defined by:

$$p_{i_1,i_2} = \frac{N_{i_1}(s)}{D_{i_2}(s)}, \tag{2.15}$$

with $i_1, i_2 \in \{1, 2, 3, 4\}$. □

Definition 2.13 (*Associated Close-Loop Polynomials*).   For the feedback loop in Fig. 2.4, we associate a close-loop polynomial

$$p_{i_1,i_2} = k(s-z)N_{i_1}(s) + (s-p)D_{i_2}(s), \tag{2.16}$$

with each Kharitonov plant $p_{i_1,i_2}(s)$. □

For instance, for

$$p(s, q, r) = \frac{[4.5, 5.5]\,s^3 + [3.5, 4.5]\,s^2 + [2.5, 3.5]\,s + [6.5, 7.5]}{s^3 + [4.5, 5.5]\,s^2 + [5.5, 6.5]\,s + [7.5, 8.5]} \tag{2.17}$$

and $C(s) = \frac{1}{s+1}$,

$$p_{2,3}(s) = \frac{4.5s^3 + 3.5s^2 + 3.5s + 7.5}{s^3 + 4.5s^2 + 5.5s + 8.5}, \tag{2.18}$$

and associated close loop polynomial is:

$$p_{cl}(s) = s^4 + 10s^3 + 13.5s^2 + 17.5s + 16. \tag{2.19}$$

Theorem 2.14   (16-Plant Theorem).   *Consider the strictly proper interval plant family with first order compensator $C(s) = k\frac{s-z}{s-p}$. $C(s)$ robustly stabilize P if and only if it stabilizes each of Kharitonov's 16 plants.* □

Following pseudo-code shows the design procedure.

for $k = k_{min} : k_{step} : k_{max}$
    for $z = z_{min} : z_{step} : z_{max}$
        for $p = p_{min} : p_{step} : p_{max}$
            if 16 Kharitonov plants are Hurwitz, put the k, z and p
            in the acceptable set.
        next p
    next z
next k

Here is the summary of a design-based 16-plant theorem.

(a) Associated 16 plants, i.e., $p_{i_1,i_2} = \frac{N_{i_1}(s)}{D_{i_2}(s)}$, $i_1, i_2 \in \{1, 2, 3, 4\}$ are calculated.

(b) A first-order controller, i.e., denominator degree equals to one, is selected. A 16-plant theorem is not valid for higher-order controllers.

(c) Using the Routh–Hurwitz table, a suitable range of acceptable parameters is calculated for each of the 16 plants separately.

(d) Intersection of the obtained 16 inequalities defines the acceptable range of controller parameters which robustly stabilize the interval plant.

A reasonable question may rise at this point: Using the first or second approach only the stability of a close-loop system is obtained, how is the required performance, i.e., phase margin and gain margin, satisfied?

When the required performance is given in terms of gain/phase margin, the designer can add a fictitious block $(Ae^{-j\varphi})$ to the block diagram of Fig. 2.5 and design $C(s)$ such that it stabilizes the interval plant and the new block, i.e., $C(s)$ stabilize $P(s) \times Ae^{-j\varphi}$. Details of method can be found in Tan et al. [2006]. For a numerical example, see Yeroglu and Tan [2008].

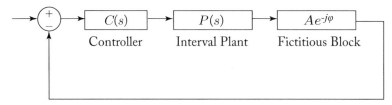

Figure 2.5: Feedback control with fictitious block $Ae^{-j\varphi}$.

2.4   CASE STUDY: ROBUST CONTROL OF A POSITION CONTROL SYSTEM

Figure 2.6 shows the block diagram of a position control system. A DC motor is used as an actuator.

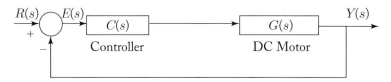

Figure 2.6: Block diagram of a position control system.

The DC motor transfer function is given by:

$$G(s) = \frac{K_m}{s(Js+b)(L_f s + R_f)} = \frac{K_m}{L_f J s^3 + (bL_f + JR_f) s^2 + bR_f s}, \qquad (2.20)$$

where R_f, L_f, K_m, b, and J show the armature resistance and inductance, motor torque constant, coefficient of viscouse friction and inertia of the rotor, respectively. Consider a DC motor with the following values in Table 2.2.

Table 2.2: Specification of the DC motor in Fig. 2.6

	Nominal value	Variation (%)	Variation
R_f	1.5 Ω	10 %	[1.35 Ω,1.65 Ω]
L_f	15 mH	20 %	[12 mH,18 mH]
K_m	$60 \times 10^{-3} \frac{N.m}{A}$	10 %	$[54 \times 10^{-3} \frac{N.m}{A}, 66 \times 10^{-3} \frac{N.m}{A}]$
b	2×10^{-3} N.m.s	15 %	$[1.7 \times 10^{-3}$ N.m.s, 2.3×10^{-3} N.m.s]
J	1.7×10^{-3} kg.m^2	40 %	$[1.02 \times 10^{-3}$ kg.m^2, 2.38×10^{-3} kg.m^2]

The nominal transfer function is:

$$G_0\left(s\right) = \frac{0.06}{2.55 \times 10^{-5}s^3 + 0.00258s^2 + 0.003s}. \tag{2.21}$$

The lower and upper bounds of coefficients is shown in Table 2.3.

Table 2.3: Lower and upper bounds of coefficients

	Lower Bound	Upper Bound
K_m	54×10^{-3}	66×10^{-3}
$L_f J$	12.24×10^{-5}	42.83×10^{-5}
$bL_f + JR_f$	0.0014	0.004
bR_f	0.0023	0.0038

Interval plant model of the DC motor is:

$$G\left(s\right) = \frac{q_0}{p_3 s^3 + p_2 s^2 + p_1 s + p_0}, \tag{2.22}$$

where

$$q_0 \in \left[54 \times 10^{-3}, 66 \times 10^{-3}\right],$$
$$p_3 \in \left[12.24 \times 10^{-3}, 42.83 \times 10^{-3}\right],$$
$$p_2 \in [0.0014, 0.004],$$
$$p_1 \in [0.0023, 0.0038],$$
$$p_0 \in [0, 0].$$

Since the numerator is a simple constant (it has only two different representations namely q_0^- and q_0^+), the number of Kharitonov plants associated with Equation (2.22) reduces to 8. Table 2.4 shows the Kharitonov plants.

We want to design a robust PI controller for the block diagram shown in Fig. 2.4. The following program implements the pseudo code shown in Section 2.3.2. Figure 2.7, shows the founded acceptable proportional and integral gains.

Table 2.4: The Kharitonov plants associated with Equation (2.22)

$G_1(s) =$	$\dfrac{0.054}{42.83 \times 10^{-5}\,s^3 + 0.004s^2 + 0.0023s}$
$G_2(s) =$	$\dfrac{0.054}{12.24 \times 10^{-5}\,s^3 + 0.004s^2 + 0.0038s}$
$G_3(s) =$	$\dfrac{0.054}{42.83 \times 10^{-5}\,s^3 + 0.0014s^2 + 0.0023s}$
$G_4(s) =$	$\dfrac{0.054}{12.24 \times 10^{-5}\,s^3 + 0.0014s^2 + 0.0038s}$
$G_5(s) =$	$\dfrac{0.066}{42.83 \times 10^{-5}\,s^3 + 0.004s^2 + 0.0023s}$
$G_6(s) =$	$\dfrac{0.066}{12.24 \times 10^{-5}\,s^3 + 0.004s^2 + 0.0038s}$
$G_7(s) =$	$\dfrac{0.066}{42.83 \times 10^{-5}\,s^3 + 0.0014s^2 + 0.0023s}$
$G_8(s) =$	$\dfrac{0.066}{12.24 \times 10^{-5}\,s^3 + 0.0014s^2 + 0.0038s}$

```
%This program finds the proportional gain and integral gain
%values to stabilize all the 8 Kharitonov plant
clc

%Search interval
Kpmin=0;
Kpstep=.001;
Kpmax=.16;

Kimin=0;
Kistep=.0005;
Kimax=.05;

%Kharitonov plants
G1=tf(.054,[42.83e-5 .004 .0023 0]);
G2=tf(.054,[12.24e-5 .004 .0038 0]);
G3=tf(.054,[42.83e-5 .0014 .0023 0]);
```

```
G4=tf(.054,[12.24e-5 .0014 .0038 0]);

G5=tf(.066,[42.83e-5 .004 .0023 0]);
G6=tf(.066,[12.24e-5 .004 .0038 0]);
G7=tf(.066,[42.83e-5 .0014 .0023 0]);
G8=tf(.066,[12.24e-5 .0014 .0038 0]);

%Acceptable values of proportional gain and integral gains are
%stored in variables KP and KI respectively.
KP=0;
KI=0;

%n is used to show the progress of the loop to the user.
%N is the total number of iterations
n=0;
N=length(Kpmin:Kpstep:Kpmax)*length(Kimin:Kistep:Kimax);

for Kp=Kpmin:Kpstep:Kpmax
    for Ki=Kimin:Kistep:Kimax
        n=n+1;
        disp('percentage of work done: ')
        disp(100*n/N)

        C=tf([Kp Ki],[1 0]);
        P11=pole(feedback(C*G1,1));
        P12=pole(feedback(C*G2,1));
        P13=pole(feedback(C*G3,1));
        P14=pole(feedback(C*G4,1));
        P21=pole(feedback(C*G5,1));
        P22=pole(feedback(C*G6,1));
        P23=pole(feedback(C*G7,1));
        P24=pole(feedback(C*G8,1));

NumberOfUnstablePoles=sum(real(P11)>0)+sum(real(P12)>0)...
   +sum(real(P13)>0)+sum(real(P14)>0)+sum(real(P21)>0)...
   +sum(real(P22)>0)+sum(real(P23)>0)+sum(real(P24)>0);
        if NumberOfUnstablePoles==0
            KP=[KP;Kp];
            KI=[KI;Ki];
```

```
            end
        end
end
plot(KP,KI,'.'),grid minor
xlabel('Kp')
ylabel('Ki')
```

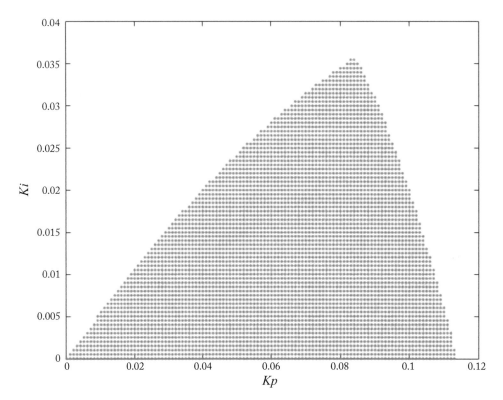

Figure 2.7: Acceptable proportional and integral gains.

We can obtain the same result using another technique. We calculate the stabilizing region for each of the Kharitonov plants separately and intersect the obtained results to obtain the stabilizing region.

A transfer function $G(s) = \frac{N(s)}{D(s)}$ can be transformed into the following form [Bhattacharyya et al., 1995]:

$$G(j\omega) = \frac{N(j\omega)}{D(j\omega)} = \frac{N_e\left(-\omega^2\right) + j\omega \cdot N_o\left(-\omega^2\right)}{D_e\left(-\omega^2\right) + j\omega \cdot D_o\left(-\omega^2\right)}, \tag{2.23}$$

where N_e, N_o, D_e, and D_o shows the even part of numerator, odd part of numerator, even part of denominator and odd part of denominator, respectively.

The characteristic equation for the block diagram shown in Fig. 2.6 is $\Delta(s) = 1 + G(s) \cdot C(s)$. Variable s is replaced with $j\omega$ to obtain the proportional and integral gains $(K_p, K_i > 0)$ which stabilize the loop:

$$\Delta(j\omega) = 1 + \frac{N(j\omega)}{D(j\omega)} C(j\omega) = 1 + \frac{N_e(-\omega^2) + j\omega \cdot N_o(-\omega^2)}{D_e(-\omega^2) + j\omega \cdot D_o(-\omega^2)} \cdot \frac{K_i + jK_p\omega}{j\omega} = 0. \quad (2.24)$$

$\Delta(j\omega)$ has two parts: a real part $(\Delta_R(j\omega))$ and an imaginary part $(\Delta_I(j\omega))$. $\Delta(j\omega) = 0$ means $\Delta_R(j\omega) = \Delta_I(j\omega) = 0$. This leads to the following system of equations:

$$\begin{cases} (-\omega^2 N_o(-\omega^2)) K_p + N_e(-\omega^2) K_i = \omega^2 D_o(-\omega^2) \\ N_e(-\omega^2) K_p + N_o(-\omega^2) K_i = -D_e(-\omega^2). \end{cases} \quad (2.25)$$

Solution of the obtained system is:

$$\begin{aligned} K_p &= -\frac{\omega^2 N_o(-\omega^2) D_o(-\omega^2) + N_e(-\omega^2) D_e(-\omega^2)}{N_e^2(-\omega^2) + \omega^2 N_o^2(-\omega^2)} \\ K_i &= \frac{\omega^2(N_e(-\omega^2) D_o(-\omega^2) - N_o(-\omega^2) D_e(-\omega^2))}{N_e^2(-\omega^2) + \omega^2 N_o^2(-\omega^2)}. \end{aligned} \quad (2.26)$$

Studying a numeric example is quite helpful.

For example, for

$$G_1(s) = \frac{0.054}{42.83 \times 10^{-5} s^3 + 0.004 s^2 + 0.0023 s},$$

$$G_1(j\omega) = \frac{0.054}{-0.004\omega^2 + j\omega(-42.83 \times 10^{-5}\omega^2 + 0.0023)},$$

$$\begin{aligned} N_e(-\omega^2) &= 0.054, \\ N_o(-\omega^2) &= 0, \\ D_e(-\omega^2) &= -0.004\omega^2, \\ D_o(-\omega^2) &= -42.83 \times 10^{-5}\omega^2 + 0.0023, \end{aligned}$$

$$K_p = -\frac{0.054 \times -0.004\omega^2}{0.054^2} = 0.0741\omega^2$$

and

$$K_i = \frac{\omega^2(0.054 \times (-42.83 \times 10^{-5}\omega^2 + 0.0023))}{0.054^2} = 0.0079\omega^4 + 0.0426\omega^2.$$

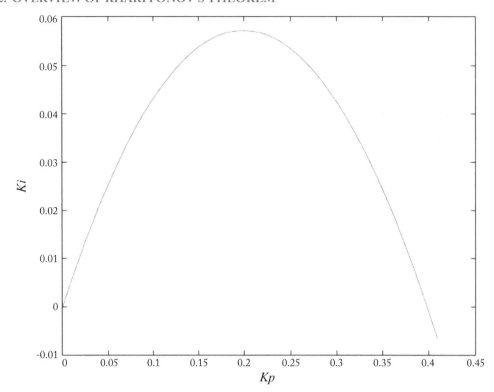

Figure 2.8: $(K_p, K_i) = (0.0741\omega^2, 0.0079\omega^4 + 0.0426\omega^2)$ tupple for $0 < \omega < 2.35\ \frac{\text{Rad}}{\text{s}}$.

Figure 2.8, shows the $(K_p, K_i) = (0.0741\omega^2, 0.0079\omega^4 + 0.0426\omega^2)$ tupple for $0 < \omega < 2.35\frac{\text{Rad}}{\text{s}}$. Figure 2.8 is produced with the aid of the following program.

```
%This program finds the acceptable range of Kp and Ki
%to stabilize G1
clc

%Desired frequency range
omega_min=0;
omega_step=.01;
omega_max=2.35;

%plant(G(s)) and its even(with subscrip e),
%odd(with subscrip o), decomposition.
%
```

```
%            N(jw)        Ne(-w^2)+jw.No(-w^2)
% G(jw)= ------- = ---------------------
%            D(jw)        De(-w^2)+jw.Do(-w^2)
%
%            -w^2.No.Do-Ne.De
% Kp = ------------------
%            Ne^2+w^2.No^2
%
%            w^2.(Ne.Do-No.De)
% Ki = ------------------
%            Ne^2+w^2.No^2


syms w
N1=.054;
D1=[42.83e-5 .004 .0023 0];
G1=tf(N1,D1);
%even and odd decomposition of plant
Ne_G1=.054;
No_G1=0;
De_G1=-0.004*w^2;
Do_G1=0.0023-42.83e-5*w^2;
%Kp and Ki
Kp_G1=-(w^2*No_G11*Do_G11+Ne_G11*De_G11)/(Ne_G11^2+w^2*
    No_G11^2);
Ki_G1=w^2*(Ne_G11*Do_G11-No_G11*De_G11)/(Ne_G11^2+w^2*
    No_G11^2);

%KP_G1 is the acceptable proportional gains which stabilize
%the loop
%KI_G1 is the acceptable integral gains which stabilize the loop
%Next to lines are initializations
KP_G1=0;
KI_G1=0;
for omega=[omega_min:omega_step:omega_max]
    KP_G1=[KP_G1;subs(Kp_G1,w,omega)];
    KI_G1=[KI_G1;subs(Ki_G1,w,omega)];
end
%polotting
plot(KP_G1,KI_G1),hold on
```

```
grid minor
xlabel('Kp')
ylabel('Ki')
```

The curve shown in Fig. 2.8 divides the plane into two regions. Points that lie between the curve and K_p axis (see Fig. 2.9) are the stable region (i.e., closed-loop poles lie in LHP).

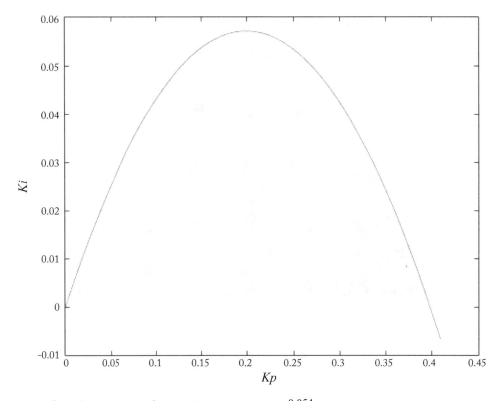

Figure 2.9: Stabilzing region for $G_1(s) = \frac{0.054}{42.83 \times 10^{-5} s^3 + 0.004 s^2 + 0.0023 s}$.

Figures 2.10 and 2.11 show the stable region for eight of Kharitonov's plants. The shaded region shows the intersection of all Kharitonov's plants. Obtained result is the same as Fig. 2.7. Figure 2.10 is produced with the aid of the following program.

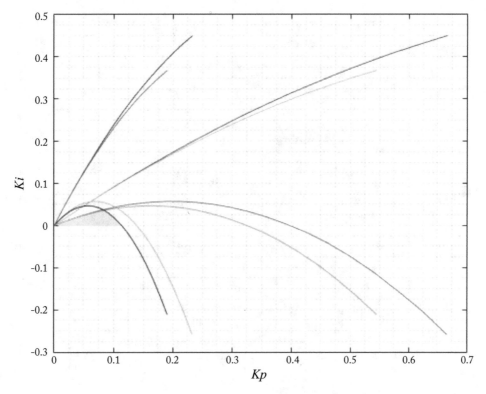

Figure 2.10: Coefficients selected from the shaded region will stabilize eight of Kharitonov's plants.

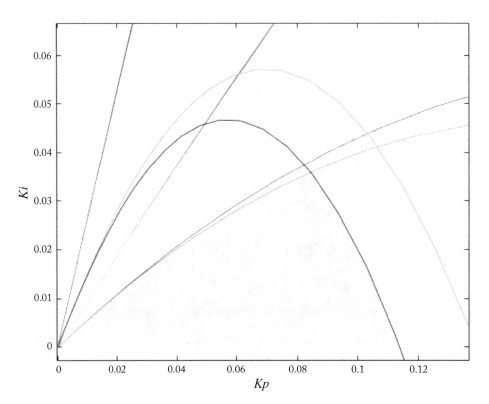

Figure 2.11: A closer look to the stabilizing region.

```
%This program stabilizes the 8 Kharitonov plants

clc

%Desired frequency range
omega_min=0;
omega_step=.1;
omega_max=3;
%system parameters
q0min=54e-3;
q0max=66e-3;

p0min=0;
p0max=0;

p1min=2.3e-3;
p1max=3.8e-3;

p2min=1.4e-3;
p2max=4e-3;

p3min=12.24e-5;
p3max=42.83e-5;
%Kharitonov's polynomials of Numerator(N) and Denominator(D)
N1=q0min;
N2=q0max;

D1=[p3max p2max p1min p0min];
D2=[p3min p2max p1max p0min];
D3=[p3max p2min p1min p0max];
D4=[p3min p2min p1max p0max];

%8 Kharitonov plants and their even (with subscrip e),
%odd (with subscrip o), decomposition.
%
%              N(jw)      Ne(-w^2)+jw.No(-w^2)
% G(jw)=  -------  =  ----------------------
%              D(jw)      De(-w^2)+jw.Do(-w^2)
```

```
%
%              -w^2.No.Do-Ne.De
% Kp = -------------------
%             Ne^2+w^2.No^2
%
%             w^2.(Ne.Do-No.De)
% Ki = -------------------
%             Ne^2+w^2.No^2
syms w

G1=tf(N1,D1);
Ne_G1=q0min;
No_G1=0;
De_G1=p0min-p2max*w^2;
Do_G1=p1min-p3max*w^2;
Kp_G1=-(w^2*No_G1*Do_G1+Ne_G1*De_G1)/(Ne_G1^2+w^2*No_G1^2);
Ki_G1=w^2*(Ne_G1*Do_G1-No_G1*De_G1)/(Ne_G1^2+w^2*No_G1^2);

G2=tf(N1,D2);
Ne_G2=q0min;
No_G2=0;
De_G2=p0min-p2max*w^2;
Do_G2=p1max-p3min*w^2;
Kp_G2=-(w^2*No_G2*Do_G2+Ne_G2*De_G2)/(Ne_G2^2+w^2*No_G2^2);
Ki_G2=w^2*(Ne_G2*Do_G2-No_G2*De_G2)/(Ne_G2^2+w^2*No_G2^2);

G3=tf(N1,D3);
Ne_G3=q0min;
No_G3=0;
De_G3=p0max-p2min*w^2;
Do_G3=p1min-p3max*w^2;
Kp_G3=-(w^2*No_G3*Do_G3+Ne_G3*De_G3)/(Ne_G3^2+w^2*No_G3^2);
Ki_G3=w^2*(Ne_G3*Do_G3-No_G3*De_G3)/(Ne_G3^2+w^2*No_G3^2);

G4=tf(N1,D4);
Ne_G4=q0min;
No_G4=0;
De_G4=p0max-p2min*w^2;
Do_G4=p1max-p3min*w^2;
```

```
Kp_G4=-(w^2*No_G4*Do_G4+Ne_G4*De_G4)/(Ne_G4^2+w^2*No_G4^2);
Ki_G4=w^2*(Ne_G4*Do_G4-No_G4*De_G4)/(Ne_G4^2+w^2*No_G4^2);

G5=tf(N2,D1);
Ne_G5=q0max;
No_G5=0;
De_G5=p0min-p2max*w^2;
Do_G5=p1min-p3max*w^2;
Kp_G5=-(w^2*No_G5*Do_G5+Ne_G5*De_G5)/(Ne_G5^2+w^2*No_G5^2);
Ki_G5=w^2*(Ne_G5*Do_G5-No_G5*De_G5)/(Ne_G5^2+w^2*No_G5^2);

G6=tf(N2,D2);
Ne_G6=q0max;
No_G6=0;
De_G6=p0min-p2max*w^2;
Do_G6=p1max-p3min*w^2;
Kp_G6=-(w^2*No_G6*Do_G6+Ne_G6*De_G6)/(Ne_G6^2+w^2*No_G6^2);
Ki_G6=w^2*(Ne_G6*Do_G6-No_G6*De_G6)/(Ne_G6^2+w^2*No_G6^2);

G7=tf(N2,D3);
Ne_G7=q0max;
No_G7=0;
De_G7=p0max-p2min*w^2;
Do_G7=p1min-p3max*w^2;
Kp_G7=-(w^2*No_G7*Do_G7+Ne_G7*De_G7)/(Ne_G7^2+w^2*No_G7^2);
Ki_G7=w^2*(Ne_G7*Do_G7-No_G7*De_G7)/(Ne_G7^2+w^2*No_G7^2);

G8=tf(N2,D4);
Ne_G8=q0max;
No_G8=0;
De_G8=p0max-p2min*w^2;
Do_G8=p1max-p3min*w^2;
Kp_G8=-(w^2*No_G8*Do_G8+Ne_G8*De_G8)/(Ne_G8^2+w^2*No_G8^2);
Ki_G8=w^2*(Ne_G8*Do_G8-No_G8*De_G8)/(Ne_G8^2+w^2*No_G8^2);

%poltting the obtaibed curves.
%KP_Gi=0 and KI_Gi=0 (i=1..8)are initializations
KP_G1=0;
KP_G2=0;
```

```
KP_G3=0;
KP_G4=0;
KP_G5=0;
KP_G6=0;
KP_G7=0;
KP_G8=0;

KI_G1=0;
KI_G2=0;
KI_G3=0;
KI_G4=0;
KI_G5=0;
KI_G6=0;
KI_G7=0;
KI_G8=0;

for omega=[omega_min:omega_step:omega_max]
    KP_G1=[KP_G1;subs(Kp_G1,w,omega)];
    KP_G2=[KP_G2;subs(Kp_G2,w,omega)];
    KP_G3=[KP_G3;subs(Kp_G3,w,omega)];
    KP_G4=[KP_G4;subs(Kp_G4,w,omega)];

    KP_G5=[KP_G5;subs(Kp_G5,w,omega)];
    KP_G6=[KP_G6;subs(Kp_G6,w,omega)];
    KP_G7=[KP_G7;subs(Kp_G7,w,omega)];
    KP_G8=[KP_G8;subs(Kp_G8,w,omega)];

    KI_G1=[KI_G1;subs(Ki_G1,w,omega)];
    KI_G2=[KI_G2;subs(Ki_G2,w,omega)];
    KI_G3=[KI_G3;subs(Ki_G3,w,omega)];
    KI_G4=[KI_G4;subs(Ki_G4,w,omega)];

    KI_G5=[KI_G5;subs(Ki_G5,w,omega)];
    KI_G6=[KI_G6;subs(Ki_G6,w,omega)];
    KI_G7=[KI_G7;subs(Ki_G7,w,omega)];
    KI_G8=[KI_G8;subs(Ki_G8,w,omega)];
end

plot(KP_G1,KI_G1),hold on
```

```
plot(KP_G2,KI_G2),hold on
plot(KP_G3,KI_G3),hold on
plot(KP_G4,KI_G4),hold on
plot(KP_G5,KI_G5),hold on
plot(KP_G6,KI_G6),hold on
plot(KP_G7,KI_G7),hold on
plot(KP_G8,KI_G8),hold on
grid minor
```

Figure 2.12, shows the step response of the uncertain system for $K_p = 0.06$ and $K_i = 0.01$ (selected gains lie inside the shaded region). This figure is produced with the aid of the following program.

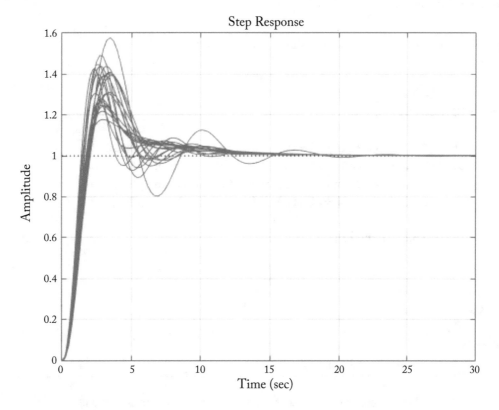

Figure 2.12: Closed-loop step response of the system with $K_p = 0.06$ and $K_i = 0.01$.

```
clc

%uncertain DC motor plant
q0min=54e-3;
q0max=66e-3;

p0min=0;
p0max=0;
p0=0;

p1min=2.3e-3;
p1max=3.8e-3;

p2min=1.4e-3;
p2max=4e-3;

p3min=12.24e-5;
p3max=42.83e-5;

q0=ureal('q0',.5*(q0min+q0max),'Range',[q0min q0max]);
p1=ureal('p1',.5*(p1min+p1max),'Range',[p1min p1max]);
p2=ureal('p2',.5*(p2min+p2max),'Range',[p2min p2max]);
p3=ureal('p3',.5*(p3min+p3max),'Range',[p3min p3max]);

G=tf(q0,[p3 p2 p1 p0]);

%Controller
Kp=.06;
Ki=.01;
s=tf('s');
C=Kp+Ki/s;
step(feedback(G*C,1)), grid minor
```

We can design the controller to satisfy some performance criterion as well. We add a ficti-tious block $Ae^{-j\phi}$ to the block diagram shown in Fig. 2.5 to obtain a certain level of gain/phase margin (see Fig. 2.13). A and φ are the desired gain and phase margin, respectively.

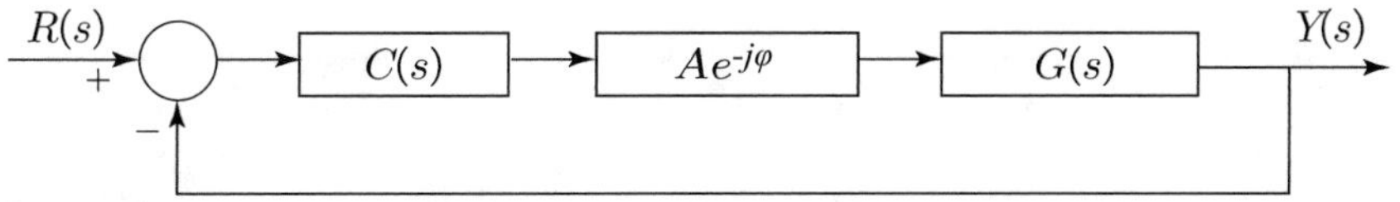

Figure 2.13: Block diagram of control system with the fictious block $Ae^{-j\phi}$.

According to the Fig. 2.13,

$$\Delta(s) = 1 + G(s) \cdot C(s) \cdot Ae^{-j\phi} = 1 + G(s) \cdot C(s) \cdot (A\cos(\varphi) - jA\sin(\varphi))$$

$$\Delta(j\omega) = 1 + \frac{N(j\omega)}{D(j\omega)} \cdot \frac{K_i + jK_p\omega}{j\omega} \cdot (A\cos(\varphi) - jA\sin(\varphi))$$

$$= 1 + \frac{N_e + j\omega \cdot N_o}{D_e + j\omega \cdot D_o} \cdot \frac{K_i + jK_p\omega}{j\omega} \cdot (A\cos(\varphi) - jA\sin(\varphi))$$

$$= \Delta_R(j\omega) + \Delta_I(j\omega),$$

(2.27)

N_e, N_o, D_e, and D_o show $N_e(-\omega^2)$, $N_o(-\omega^2)$, $D_e(-\omega^2)$, and $D_o(-\omega^2)$, respectively. Solution of $\Delta_R(j\omega) = \Delta_I(j\omega) = 0$ is

$$K_p = \frac{\sin(\varphi)\,\omega\,(N_e D_o - N_o D_e) - \cos(\varphi)\,(N_e D_e + \omega^2 N_o D_o)}{A\,(N_e^2 + \omega^2 N_o^2)}$$

$$K_i = \frac{\sin(\varphi)\,\omega\,(N_e D_e - \omega^2 N_o D_o) + \cos(\varphi)\,\omega^2(N_e D_o - N_o D_e)}{A(N_e^2 + \omega^2 N_o^2)}.$$

(2.28)

If we substitute $A = 1$ and $\varphi = 0$ in the above equations, we obtain the previous results (Equation (2.26)).

Figure 2.14 shows the region which provides the 30° of phase margin ($A = 1$ and $\varphi = \frac{\pi}{6}$). The following program is used to find this region.

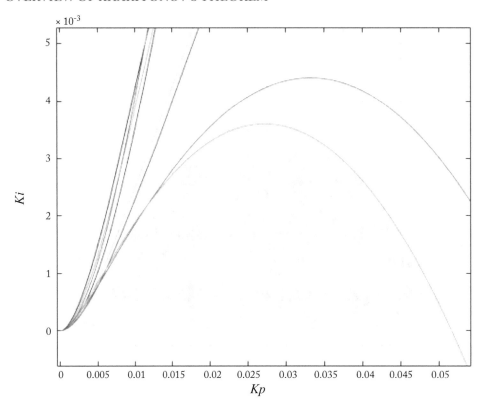

Figure 2.14: Region which provides the 30° of phase margin.

```
%This program stabilizes the 8 Kharitonov plants
%We obtain at least 30 Degrees of phase margin
clc

%Desired frequency range
omega_min=0;
omega_step=.01;
omega_max=1;

%Desired phase and gain margin
A=1;
phi=30*pi/180;

%system parameters
```

```
q0min=54e-3;
q0max=66e-3;

p0min=0;
p0max=0;

p1min=2.3e-3;
p1max=3.8e-3;

p2min=1.4e-3;
p2max=4e-3;

p3min=12.24e-5;
p3max=42.83e-5;

%Kharitonov's polynomials of Numerator(N) and Denominator(D)
N1=q0min;
N2=q0max;
D1=[p3max p2max p1min p0min];
D2=[p3min p2max p1max p0min];
D3=[p3max p2min p1min p0max];
D4=[p3min p2min p1max p0max];

%8 Kharitonov plants and their even (with subscrip e),
%odd (with subscrip o), decomposition.
%
%            N(jw)        Ne(-w^2)+jw.No(-w^2)
% G(jw)= ------- = ----------------------
%            D(jw)        De(-w^2)+jw.Do(-w^2)
%
%            sin(phi).w.(Ne.Do-No.De)-cos(phi).(Ne.De+w^2.No.Do)
% Kp = -----------------------------------------------------------
%                             A.(Ne^2+w^2.No^2)
%
%            sin(phi).w.(Ne.De+w^2.No.Do)+cos(phi).w^2.(Ne.Do-No.De)
% Ki = -----------------------------------------------------------
%                             A.(Ne^2+w^2.No^2)
syms w
G1=tf(N1,D1);
```

```
Ne_G1=q0min;
No_G1=0;
De_G1=p0min-p2max*w^2;
Do_G1=p1min-p3max*w^2;
Kp_G1=(sin(phi)*w*(Ne_G1*Do_G1-No_G1*De_G1)-cos(phi)*
    (Ne_G1*De_G1+w^2*No_G1*Do_G1))/(A*(Ne_G1^2+w^2*No_G1^2));
Ki_G1=(sin(phi)*w*(Ne_G1*De_G1+w^2*No_G1*Do_G1)+cos(phi)*
    w^2*(Ne_G1*Do_G1-No_G1*De_G1))/(A*(Ne_G1^2+w^2*No_G1^2));

G2=tf(N1,D2);
Ne_G2=q0min;
No_G2=0;
De_G2=p0min-p2max*w^2;
Do_G2=p1max-p3min*w^2;
Kp_G2=(sin(phi)*w*(Ne_G2*Do_G2-No_G2*De_G2)-cos(phi)*
    (Ne_G2*De_G2+w^2*No_G2*Do_G2))/(A*(Ne_G2^2+w^2*No_G2^2));
Ki_G2=(sin(phi)*w*(Ne_G2*De_G2+w^2*No_G2*Do_G2)+cos(phi)*
    w^2*(Ne_G2*Do_G2-No_G2*De_G2))/(A*(Ne_G2^2+w^2*No_G2^2));

G3=tf(N1,D3);
Ne_G3=q0min;
No_G3=0;
De_G3=p0max-p2min*w^2;
Do_G3=p1min-p3max*w^2;
Kp_G3=(sin(phi)*w*(Ne_G3*Do_G3-No_G3*De_G3)-cos(phi)*
    (Ne_G3*De_G3+w^2*No_G3*Do_G3))/(A*(Ne_G3^2+w^2*No_G3^2));
Ki_G3=(sin(phi)*w*(Ne_G3*De_G3+w^2*No_G3*Do_G3)+cos(phi)*
    w^2*(Ne_G3*Do_G3-No_G3*De_G3))/(A*(Ne_G3^2+w^2*No_G3^2));

G4=tf(N1,D4);
Ne_G4=q0min;
No_G4=0;
De_G4=p0max-p2min*w^2;
Do_G4=p1max-p3min*w^2;
Kp_G4=(sin(phi)*w*(Ne_G4*Do_G4-No_G4*De_G4)-cos(phi)*
    (Ne_G4*De_G4+w^2*No_G4*Do_G4))/(A*(Ne_G4^2+w^2*No_G4^2));
Ki_G4=(sin(phi)*w*(Ne_G4*De_G4+w^2*No_G4*Do_G4)+cos(phi)*
    w^2*(Ne_G4*Do_G4-No_G4*De_G4))/(A*(Ne_G4^2+w^2*No_G4^2));
```

```
G5=tf(N2,D1);
Ne_G5=q0max;
No_G5=0;
De_G5=p0min-p2max*w^2;
Do_G5=p1min-p3max*w^2;
Kp_G5=(sin(phi)*w*(Ne_G5*Do_G5-No_G5*De_G5)-cos(phi)*
    (Ne_G5*De_G5+w^2*No_G5*Do_G5))/(A*(Ne_G5^2+w^2*No_G5^2));
Ki_G5=(sin(phi)*w*(Ne_G5*De_G5+w^2*No_G5*Do_G5)+cos(phi)*
    w^2*(Ne_G5*Do_G5-No_G5*De_G5))/(A*(Ne_G5^2+w^2*No_G5^2));

G6=tf(N2,D2);
Ne_G6=q0max;
No_G6=0;
De_G6=p0min-p2max*w^2;
Do_G6=p1max-p3min*w^2;
Kp_G6=(sin(phi)*w*(Ne_G6*Do_G6-No_G6*De_G6)-cos(phi)*
    (Ne_G6*De_G6+w^2*No_G6*Do_G6))/(A*(Ne_G6^2+w^2*No_G6^2));
Ki_G6=(sin(phi)*w*(Ne_G6*De_G6+w^2*No_G6*Do_G6)+cos(phi)*
    w^2*(Ne_G6*Do_G6-No_G6*De_G6))/(A*(Ne_G6^2+w^2*No_G6^2));

G7=tf(N2,D3);
Ne_G7=q0max;
No_G7=0;
De_G7=p0max-p2min*w^2;
Do_G7=p1min-p3max*w^2;
Kp_G7=(sin(phi)*w*(Ne_G7*Do_G7-No_G7*De_G7)-cos(phi)*
    (Ne_G7*De_G7+w^2*No_G7*Do_G7))/(A*(Ne_G7^2+w^2*No_G7^2));
Ki_G7=(sin(phi)*w*(Ne_G7*De_G7+w^2*No_G7*Do_G7)+cos(phi)*
    w^2*(Ne_G7*Do_G7-No_G7*De_G7))/(A*(Ne_G7^2+w^2*No_G7^2));

G8=tf(N2,D4);
Ne_G8=q0max;
No_G8=0;
De_G8=p0max-p2min*w^2;
Do_G8=p1max-p3min*w^2;
Kp_G8=(sin(phi)*w*(Ne_G8*Do_G8-No_G8*De_G8)-cos(phi)*
    (Ne_G8*De_G8+w^2*No_G8*Do_G8))/(A*(Ne_G8^2+w^2*No_G8^2));
Ki_G8=(sin(phi)*w*(Ne_G8*De_G8+w^2*No_G8*Do_G8)+cos(phi)*
    w^2*(Ne_G8*Do_G8-No_G8*De_G8))/(A*(Ne_G8^2+w^2*No_G8^2));
```

```
%poltting the obtaibed curves.
%KP_Gij=0 and KI_Gij=0 (i,j=1..4)are initializations
KP_G1=0;
KP_G2=0;
KP_G3=0;
KP_G4=0;
KP_G5=0;
KP_G6=0;
KP_G7=0;
KP_G8=0;

KI_G1=0;
KI_G2=0;
KI_G3=0;
KI_G4=0;
KI_G5=0;
KI_G6=0;
KI_G7=0;
KI_G8=0;
for omega=[omega_min:omega_step:omega_max]
    KP_G1=[KP_G1;subs(Kp_G1,w,omega)];
    KP_G2=[KP_G2;subs(Kp_G2,w,omega)];
    KP_G3=[KP_G3;subs(Kp_G3,w,omega)];
    KP_G4=[KP_G4;subs(Kp_G4,w,omega)];

    KP_G5=[KP_G5;subs(Kp_G5,w,omega)];
    KP_G6=[KP_G6;subs(Kp_G6,w,omega)];
    KP_G7=[KP_G7;subs(Kp_G7,w,omega)];
    KP_G8=[KP_G8;subs(Kp_G8,w,omega)];

    KI_G1=[KI_G1;subs(Ki_G1,w,omega)];
    KI_G2=[KI_G2;subs(Ki_G2,w,omega)];
    KI_G3=[KI_G3;subs(Ki_G3,w,omega)];
    KI_G4=[KI_G4;subs(Ki_G4,w,omega)];

    KI_G5=[KI_G5;subs(Ki_G5,w,omega)];
    KI_G6=[KI_G6;subs(Ki_G6,w,omega)];
    KI_G7=[KI_G7;subs(Ki_G7,w,omega)];
```

```
    KI_G8=[KI_G8;subs(Ki_G8,w,omega)];
end

plot(KP_G1,KI_G1),hold on
plot(KP_G2,KI_G2),hold on
plot(KP_G3,KI_G3),hold on
plot(KP_G4,KI_G4),hold on
plot(KP_G5,KI_G5),hold on
plot(KP_G6,KI_G6),hold on
plot(KP_G7,KI_G7),hold on
plot(KP_G8,KI_G8),hold on
grid minor
```

Figure 2.15 shows the step response of the uncertain system for $K_p = 0.045$ and $K_i = 0.002$ (selected gains lie inside the shaded region shown in Fig. 2.14).

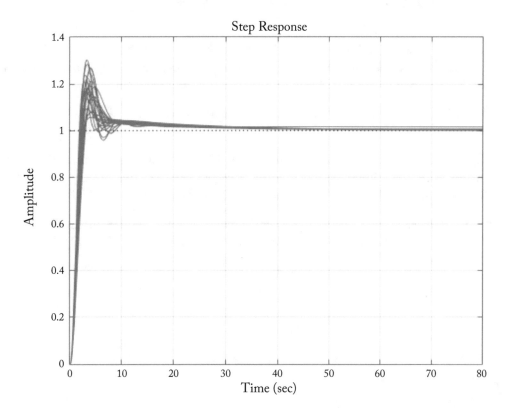

Figure 2.15: Step response of closed-loop control system with $K_p = 0.045$ and $K_i = 0.002$.

2.5   CONCLUSION

In this chapter we introduced Kharitonov's theorem. We showed how it can be used as a design tool. In the next chapter we will use the introduced methods to design robust controllers for DC-DC converters.

REFERENCES

Barmish, B. R. Kharitonov's theorem and its extensions and applications: An introduction. *26th IEEE Conference on Decision and Control*, pp. 2060–2061, Los Angeles, CA, 1987. DOI: 10.1109/cdc.1987.272917.

Belanger, P. R. *Control Engineering: A Modern Approach*, Oxford University Press, 2005.

Bhattacharyya, S. P., Chapellat, H., and Keel, L. H. *Robust Control: The Parametric Approach*, Prentice Hall, 1995. DOI: 10.1016/b978-0-08-042230-5.50016-5. 50

Dorf, R. C. and Bishop, R. H. *Modern Control Systems*, Prentice Hall, 2001. DOI: 10.1109/tsmc.1981.4308749.

Hollot, C. V., Kraus, F. J., Tempo, R., and Barmish, B. R. Extreme point results for robust stabilization of interval plants with first order compensators. *American Control Conference*, pp. 2533–2538, San Diego, CA, 1990. DOI: 10.23919/acc.1990.4791182.

Ross Barmish, B. *New Tools for Robustness of Linear Systems*, Macmillan Publishing Company, 1994.

Tan, N., Kaya, I., Yeroglu, C., and Atherton, D. P. Computation of stabilizing PI and PID controllers using the stability boundary locus. *Energy Conversion and Management*, vol. 47, pp. 3045–3058, 2006. DOI: 10.1016/j.enconman.2006.03.022. 45

Yeroglu, C. and Tan, N. Design of robust PI controller for vehicle suspension system. *Journal of Electrical Engineering and Technology*, vol. 3, no. 1, pp. 135–142, 2008. DOI: 10.5370/jeet.2008.3.1.135. 45

CHAPTER 3

Controller Design for DC-DC Converters Using Kharitonov's Theorem

3.1 INTRODUCTION

In this chapter we show how Kharitonov's theorem can be used to design controllers for DC-DC converters.

The design procedure is shown with three examples. For the sake of simplicity, we assume that only output load changes. You can study the effect of other changes in the same way shown in this chapter.

Here is the summary of the method used in this chapter to design the controller.

- Control-to-output transfer function is extracted for different load values. Maximum and minimums of transfer function coefficients are extracted. This step gives the interval plant model of converter.

- A controller is selected. In this chapter we assumed a PI controller since it is simple and practical. Close-loop transfer function is calculated in presence of selected controller.

- Kharitonov's theorem is applied to the denominator of the obtained close-loop transfer function.

- Controller parameters are selected from the acceptable region and simulations are done.

3.2 ROBUST CONTROLLER DESIGN FOR QUADRATIC BUCK CONVERTER

Figure 3.1 shows the schematic of a quadratic buck converter.

When MOSFET M is closed $D1$, $D2$, and $D3$ are reverse biased, forward biased, and reverse biased, respectively. Figure 3.2 shows the equivalent circuit for a closed switch.

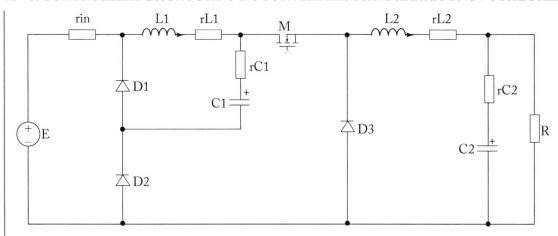

Figure 3.1: Quadratic buck converter.

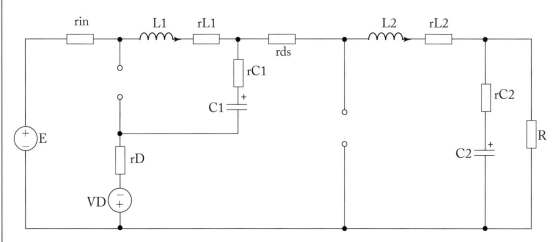

Figure 3.2: Equivalent circuit for closed MOSFET.

When MOSFET switch is closed, converter differential equations can be written as

$$L_1 \dot{i_1} = -(r_{in} + r_{L1} + r_{C1} + r_D) i_{L1} + (r_{C1} + r_D) i_{L2} - v_{C1} + E + V_D$$

$$L_2 \dot{i_2} = -\left(r_{ds} + r_{L2} + r_{C1} + r_D + \frac{r_{C2} R}{r_{C2} + R}\right) i_{L2} + (r_{C1} + r_D) i_{L1}$$

$$+ v_{C1} - \left(\frac{R}{R + r_{C2}}\right) v_{C2} - V_D \tag{3.1}$$

$$C_1 \dot{v_{C1}} = i_{L1} - i_{L2}$$

$$C_2 \dot{v_{C2}} = \frac{R}{R + r_{C2}} i_{L1} - \frac{1}{R + r_{C2}} v_{C2}.$$

When MOSFET M is opened, $D1$, $D2$, and $D3$ are forward biased, reverse biased, and reverse biased, respectively. Figure 3.3 shows the equivalent circuit for the opened switch.

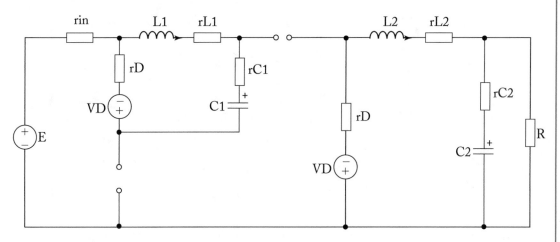

Figure 3.3: Equivalent circuit for opened MOSFET.

In this case converter dynamical equations can be written as:

$$L_1 \dot{i}_1 = -(r_{L1} + r_D)i_{L1} - v_{C1} - V_D$$

$$L_2 \dot{i}_2 = -\left(r_{L2} + r_D + \frac{r_{C2}R}{r_{C2} + R}\right)i_{L2} - \left(\frac{R}{R + r_{C2}}\right)v_{C2} - V_D$$

$$C_1 \dot{v}_{C1} = i_{L1}$$

$$C_2 \dot{v}_{C2} = \frac{R}{R + r_{C2}}i_{L2} - \frac{1}{R + r_{C2}}v_{C2}.$$

(3.2)

In both cases output (load resistor voltage) can be written as $v_{load} = \frac{R r_{C2}}{R + r_{C2}}i_{L2} + \frac{R}{R + r_{C2}}v_{C2}$.

Assume a quadratic buck converter with the following values: $E = 42$ V, $rin = 0.1$ Ω, $L1 = 350$ μH, $rL1 = 50$ mΩ, $L2 = 350$ μH, $rL2 = 30$ mΩ, $rD = 0.1$ Ω, $VD = 0.7$, $rds = 60$ mΩ, 5 $\Omega < R < 25$ Ω, $Fsw = 50$ KHz, and $D = 0.555$. D and Fsw show duty ratio and switching frequency of converter, respectively. For the aforementioned range of loads converter operates in the CCM. Figure 3.4 shows frequency response of control to output transfer function when load resistor changes from 5 Ω toward 25 Ω with 0.2 Ω steps.

The quadratic buck converters transfer function is a fourth-order transfer function (since it has four energy storage elements—two capacitors and two inductors). The general form of a proper fourth-order transfer function is:

$$\frac{\tilde{v}_o(s)}{\tilde{d}(s)} = \frac{n_3 s^3 + n_2 s^2 + n_1 s^1 + n_0}{s^4 + d_3 s^3 + d_2 s^2 + d_1 s^1 + d_0}.$$

(3.3)

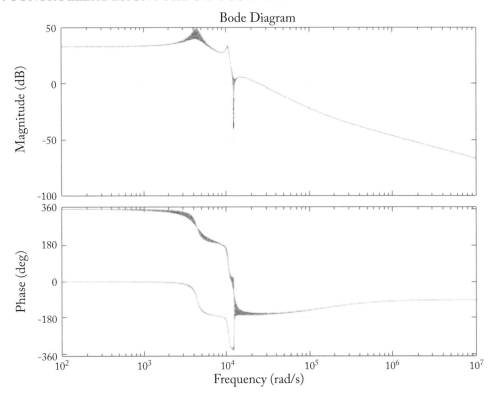

Figure 3.4: Control-to-output $\left(\frac{\tilde{v}_o(s)}{\tilde{d}(s)}\right)$ frequency response when the load changes from 5 Ω toward 25 Ω (with 0.2 Ω steps).

When the load resistor changes from 5 Ω toward 25 Ω, the transfer function coefficients changes.

In order to obtain the maximum and minimum of coefficients, the load resistor is changed from 5 Ω toward 25 Ω with 0.2 Ω steps. Control to output transfer function is calculated using SSA for each value of load resistor separately. The last step is finding the maximum and minimum values of each coefficient among the obtained 100 transfer functions $\left(\frac{25-5}{0.2} = 100\right)$.

Following code extracts the maximum and minimum of coefficients. After running the code, results shown in Table 3.1 are obtained. The code saves minimum and maximum of numerator and denominator coefficients in variables named num_min, num_max, den_min and den_min.

```
%This code is used to obtain the change
%in transfer function coefficients of
%a quadratic buck converter when output load
```

```
%changes in the [5-25] ohm range.

clc
clear all;

Rmin=5;      %Load resistor minimum value
delta_R=.2;  %Load changes is 0.2 ohm steps
Rmax=25;     %Load resistor maximum value

%NUM and DEN keeps the numerator and denominator
%coefficients. Since denominator has degree of four,
%five coefficients are required. So, DEN has five
%elements to keep d0, d1, d2, d3, d4. for instance
%d1 is the coefficient of s^1 term in the denominator.
%Numerator is of degree three and an array with four
%elements are enough.But an array of size five(the same
%size as denominator) is used. The unnecessary element
%is filled with zero.
%The NUM and DEN are initialized to zero as well.
NUM=zeros(1,5);
DEN=zeros(1,5);
N=(Rmax-Rmin)/delta_R; % number of required iteration
n=0; %counter variable

for R=Rmin:delta_R:Rmax
n=n+1;
clc
disp('Percentage of work done...')
disp(100*n/N)

%Converter parameters
VIN=42; %input source voltage
rin=.1; %input source internal resistance

L1=400e-6; %Inductor L1
rL1=.05;    %Equivalent Series Resistance of C1

L2=350e-6;
rL2=.03;
```

```
C1=33e-6; %Capacitance of capacitor C1
rC1=.05;  %Equivalent Series Resistance of C1

C2=100e-6;
rC2=.07;

rD=.01;   %Diode resistance
VD=.7;    %Diode voltage drop

rds=.06; %Drain-Source resistance

D=.555;   %Duty ratio

%Symbolic equations
syms iL1 iL2 vC1 vC2 vin vD d
%iL1 shows the current in inductor named L1
%vC1 shows the voltage of capacitor named C1

%CLOSED MOSFET EQUATIONS
M1=(-(rin+rL1+rC1+rD)*iL1+(rC1+rD)*iL2-vC1+vin+vD)/L1;
M2=(-(rds+rL2+rC1+rD+(R*rC2/(R+rC2)))*iL2+(rC1+rD)*
    iL1+vC1-(1-(rC2/(R+rC2)))*vC2-vD)/L2;
M3=(iL1-iL2)/C1;
M4=((R*iL2)-vC2)/(R+rC2)/C2;

%OPENED MOSFET EQUATIONS
M5=(-(rL1+rD)*iL1-vC1-vD)/L1;
M6=(-(rD+rL2+(R*rC2/(R+rC2)))*iL2-(R/(R+rC2)*vC2)-vD)/L2;
M7=iL1/C1;
M8=(R/(R+rC2)*iL2-1/(R+rC2)*vC2)/C2;

%AVERAGING
MA1= simplify(d*M1+(1-d)*M5);
MA2= simplify(d*M2+(1-d)*M6);
MA3= simplify(d*M3+(1-d)*M7);
MA4= simplify(d*M4+(1-d)*M8);

%DC OPERATING POINT CALCULATION
```

```
MA_DC_1=subs(MA1,[vin vD d],[VIN VD D]);
MA_DC_2=subs(MA2,[vin vD d],[VIN VD D]);
MA_DC_3=subs(MA3,[vin vD d],[VIN VD D]);
MA_DC_4=subs(MA4,[vin vD d],[VIN VD D]);

DC_SOL=
solve(MA_DC_1==0,MA_DC_2==0,MA_DC_3==0,MA_DC_4==0,
   'iL1','iL2','vC1','vC2');

IL1=eval(DC_SOL.iL1);
IL2=eval(DC_SOL.iL2);
VC1=eval(DC_SOL.vC1);
VC2=eval(DC_SOL.vC2);

%LINEARIZATION
%x=[iL1;iL2;vC1;vC2]
%u=[vin;d] where d=duty and vin= change in input voltage
A11=subs(simplify(diff(MA1,iL1)),[iL1 iL2 vC1 vC2 d vD vin],
   [IL1 IL1 VC1 VC2 D VD VIN]);
A12=subs(simplify(diff(MA1,iL2)),[iL1 iL2 vC1 vC2 d vD vin],
   [IL1 IL1 VC1 VC2 D VD VIN]);
A13=subs(simplify(diff(MA1,vC1)),[iL1 iL2 vC1 vC2 d vD vin],
   [IL1 IL1 VC1 VC2 D VD VIN]);
A14=subs(simplify(diff(MA1,vC2)),[iL1 iL2 vC1 vC2 d vD vin],
   [IL1 IL1 VC1 VC2 D VD VIN]);

A21=subs(simplify(diff(MA2,iL1)),[iL1 iL2 vC1 vC2 d vD vin],
   [IL1 IL1 VC1 VC2 D VD VIN]);
A22=subs(simplify(diff(MA2,iL2)),[iL1 iL2 vC1 vC2 d vD vin],
   [IL1 IL1 VC1 VC2 D VD VIN]);
A23=subs(simplify(diff(MA2,vC1)),[iL1 iL2 vC1 vC2 d vD vin],
   [IL1 IL1 VC1 VC2 D VD VIN]);
A24=subs(simplify(diff(MA2,vC2)),[iL1 iL2 vC1 vC2 d vD vin],
   [IL1 IL1 VC1 VC2 D VD VIN]);

A31=subs(simplify(diff(MA3,iL1)),[iL1 iL2 vC1 vC2 d vD vin],
   [IL1 IL1 VC1 VC2 D VD VIN]);
A32=subs(simplify(diff(MA3,iL2)),[iL1 iL2 vC1 vC2 d vD vin],
   [IL1 IL1 VC1 VC2 D VD VIN]);
```

```
A33=subs(simplify(diff(MA3,vC1)),[iL1 iL2 vC1 vC2 d vD vin],
   [IL1 IL1 VC1 VC2 D VD VIN]);
A34=subs(simplify(diff(MA3,vC2)),[iL1 iL2 vC1 vC2 d vD vin],
   [IL1 IL1 VC1 VC2 D VD VIN]);

A41=subs(simplify(diff(MA4,iL1)),[iL1 iL2 vC1 vC2 d vD vin],
   [IL1 IL1 VC1 VC2 D VD VIN]);
A42=subs(simplify(diff(MA4,iL2)),[iL1 iL2 vC1 vC2 d vD vin],
   [IL1 IL1 VC1 VC2 D VD VIN]);
A43=subs(simplify(diff(MA4,vC1)),[iL1 iL2 vC1 vC2 d vD vin],
   [IL1 IL1 VC1 VC2 D VD VIN]);
A44=subs(simplify(diff(MA4,vC2)),[iL1 iL2 vC1 vC2 d vD vin],
   [IL1 IL1 VC1 VC2 D VD VIN]);

A=eval([A11 A12 A13 A14;
        A21 A22 A23 A24;
        A31 A32 A33 A34;
        A41 A42 A43 A44]);

B11=subs(simplify(diff(MA1,vin)),[iL1 iL2 vC1 vC2 d vD vin],
   [IL1 IL1 VC1 VC2 D VD VIN]);
B12=subs(simplify(diff(MA1,d)),[iL1 iL2 vC1 vC2 d vD vin],
   [IL1 IL1 VC1 VC2 D VD VIN]);
B21=subs(simplify(diff(MA2,vin)),[iL1 iL2 vC1 vC2 d vD vin],
   [IL1 IL1 VC1 VC2 D VD VIN]);
B22=subs(simplify(diff(MA2,d)),[iL1 iL2 vC1 vC2 d vD vin],
   [IL1 IL1 VC1 VC2 D VD VIN]);

B31=subs(simplify(diff(MA3,vin)),[iL1 iL2 vC1 vC2 d vD vin],
   [IL1 IL1 VC1 VC2 D VD VIN]);
B32=subs(simplify(diff(MA3,d)),[iL1 iL2 vC1 vC2 d vD vin],
   [IL1 IL1 VC1 VC2 D VD VIN]);

B41=subs(simplify(diff(MA4,vin)),[iL1 iL2 vC1 vC2 d vD vin],
   [IL1 IL1 VC1 VC2 D VD VIN]);
B42=subs(simplify(diff(MA4,d)),[iL1 iL2 vC1 vC2 d vD vin],
   [IL1 IL1 VC1 VC2 D VD VIN]);

B=eval([B11 B12;
```

```
          B21 B22;
          B31 B32;
          B41 B42]);

%output equation needs no averaging since it has the
%same form for both cases(MOSFET closed and MOSFET opened).
%So we can wite the output voltage as:
%C_vR*x where x=[iL1;iL2;vC1;vC2] and C_vR is:
C_vR=[0 rC2*R/(R+rC2) 0 R/(R+rC2)];

H_vR=tf(ss(A,B,C_vR,0));
vR_d=H_vR(1,2); %vR_d is the control to output transfer function
                %since we assumed u as u=[vin;d], the transfer
                %function from second input(duty ratio) to the
                %output voltage is the desired control-to-output
                %transfer function.

[num,den] = tfdata(vR_d,'v');
NUM=[NUM;num];
DEN=[DEN;den];
end

NUM(1,:)=[]; %first row is our initialization [0 0 0 0 0],
             %so we cleaned it.
DEN(1,:)=[]; %first row is our initialization [0 0 0 0 0],
             %so we cleaned it.


%Results
num_min=min(NUM); %min of numerator coefficients when load
                  %changes in the [5-25] ohm range
num_max=max(NUM); %max of numerator coefficients when load
                  %changes in the [5-25] ohm range

den_min=min(DEN); %min of denominator coefficients when load
                  %changes in the [5-25] ohm range
den_max=max(DEN); %max of denominator coefficients when load
                  %changes in the [5-25] ohm range
%End of code
```

Table 3.1: Minimum and maximum of control-to-output transfer function coefficients

Coefficient	Minimum	Maximum
n_3	4.5778×10^3	4.6571×10^3
n_2	6.5189×10^8	6.6678×10^8
n_1	4.0827×10^{11}	9.2743×10^{11}
n_0	1.0069×10^{17}	1.0234×10^{17}
d_4	1	1
d_3	1.2452×10^3	2.8165×10^3
d_2	1.3134×10^8	1.3205×10^8
d_1	8.9629×10^{10}	2.5067×10^{11}
d_0	2.1678×10^{15}	2.1808×10^{15}

Assume a voltage control loop like that shown in Fig. 3.5. A PI controller is used to control the converter.

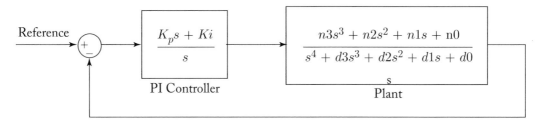

Figure 3.5: Structure of control loop.

Close-loop transfer function is:

$$\mathbf{H}(s) = \frac{\dfrac{K_p s + K_i}{s} \times \dfrac{n_3 s^3 + n_2 s^2 + n_1 s^1 + n_0}{s^4 + d_3 s^3 + d_2 s^2 + d_1 s^1 + d_0}}{1 + \dfrac{K_p s + K_i}{s} \times \dfrac{n_3 s^3 + n_2 s^2 + n_1 s^1 + n_0}{s^4 + d_3 s^3 + d_2 s^2 + d_1 s^1 + d_0}}. \qquad (3.4)$$

After simplification the close-loop denominator is obtained as:

$$s^5 + (d_3 + n_3 K_p) s^4 + (d_2 + n_2 K_p + n_3 K_i) s^3 + (d_1 + n_1 K_p + n_2 K_i) s^2$$
$$+ (d_0 + n_0 K_p + n_1 K_i) s + n_0 K_i. \qquad (3.5)$$

Minimum and maximum of these coefficients are shown in Table 3.2.

Table 3.2: Maximum and minimum of close loop coefficients (Equation (3.5))

Coefficient	Minimum	Maximum
$d_3 + n_3 K_p$	$1.2452 \times 10^3 + 4.5778 \times 10^3 K_p$	$2.8165 \times 10^3 + 4.6571 \times 10^3 K_p$
$d_2 + n_2 K_p + n_3 K_i$	$1.3134 \times 10^8 + 6.5189 \times 10^8 K_p$ $+ 4.5778 \times 10^3 K_i$	$1.3205 \times 10^8 + 6.6678 \times 10^8 K_p$ $+ 4.6571 \times 10^3 K_i$
$d_1 + n_1 K_p + n_2 K_i$	$8.9629 \times 10^{10} + 4.0827 \times$ $10^{11} K_p + 6.5189 \times 10^8 K_i$	$2.5067 \times 10^{11} + 9.2743 \times$ $10^{11} K_p + 6.6678 \times 10^8 K_i$
$d_0 + n_0 K_p + n_1 K_i$	$2.1678 \times 10^{15} + 1.0069 \times 10^{17} K_p$ $+ 4.0827 \times 10^{11} K_i$	$2.1808 \times 10^{15} + 1.0234 \times$ $10^{17} K_p + 9.2743 \times 10^{11} K_i$
$n_0 K_i$	$1.0069 \times 10^{17} K_i$	$1.0234 \times 10^{17} K_i$

We use this table to form the Kharitonov's polynomials. Since the closed-loop denominator is a fifth-order polynomial, we need to stabilize only $K_2(s)$, $K_3(s)$, and $K_4(s)$ (see Lemma 2.8).

The following code finds the suitable values of *Kp* and *Ki* which stabilize the $K_2(s)$, $K_3(s)$, and $K_4(s)$. Acceptable region found by the code is shown in Fig. 3.6. If we decrease the steps (kp_delta=.0001;), we take a better view of acceptable region (Fig. 3.7).

```
%Robust PI controller design for quadratic buck converter
clear all
clc

kp_min=0;
kp_max=.015;
kp_delta=.001; %decrese it to .0001 if you want a take
               %a better view

ki_min=0;
ki_max=10;
ki_delta=.01;

sol=[0 0]; %sol=[kp ki] keeps the acceptable gains for PI
           %controller

N=(kp_max-kp_min)/kp_delta*(ki_max-ki_min)/ki_delta;
n=0;
```

```matlab
nAcceptable=0; %nAcceptable keeps the number of acceptable
              %solutions found.

for kp=kp_min:kp_delta:kp_max
    for ki=ki_min:ki_delta:ki_max

        n=n+1;
        disp('percentage of work done')
        100*n/N
        disp('---')
        k1=roots([1 2.8165e3+4.6571e3*kp 1.3134e8+6.5189e8*kp+
            4.5778e3*ki 8.9629e10+4.0827e11*kp+6.5189e8*ki
            2.1808e15+1.0234e17*kp+9.2743e11*ki 1.0234e17*ki]);
        k1_real=real(k1);
        T1=sum(k1_real>0); %T1 keeps the number of unstable
                           %poles in k1

        k2=roots([1 2.8165e3+4.6571e3*kp 1.3134e8+6.5189e8*kp+
            4.5778e3*ki 2.5067e11+9.2743e11*kp+6.6678e8*ki
            2.1678e15+1.0069e17*kp+4.0827e11*ki 1.0234e17*ki]);
        k2_real=real(k2);
        T2=sum(k2_real>0); %T2 keeps the number of unstable
                           %poles in k2

        k3=roots([1 1.2452e3+4.5778e3*kp 1.3205e8+6.6678e8*kp+
            4.6571e3*ki 8.9629e10+4.0827e11*kp+6.5189e8*ki
            2.1808e15+1.0234e17*kp+9.2743e11*ki 1.0069e17*ki]);
        k3_real=real(k3);
        T3=sum(k3_real>0); %T3 keeps the number of unstable
                           %poles in k3

        if ((T1+T2+T3)==0)
            nAcceptable=nAcceptable+1;
            disp('kp=')
            kp
            disp('ki=')
            ki
            disp('*************')
```

```
            sol=[sol;[kp ki]];
        end
    end
end

sol(1,:)=[]; %first row is initialization[0 0]. So, we remove it.

plot(sol(:,2),sol(:,1),'.'), xlabel('Ki'), ylabel ('Kp'), grid
    minor

%End of code
```

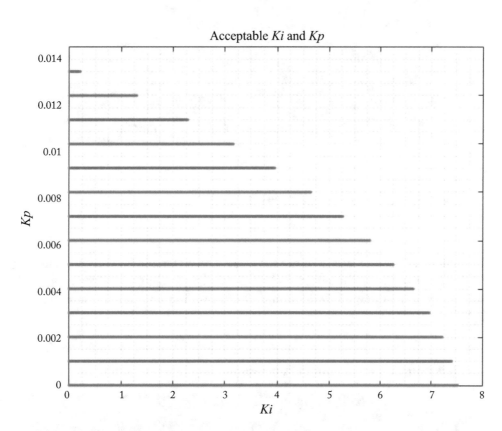

Figure 3.6: Acceptable PI coefficients (`kp_delta=.001`).

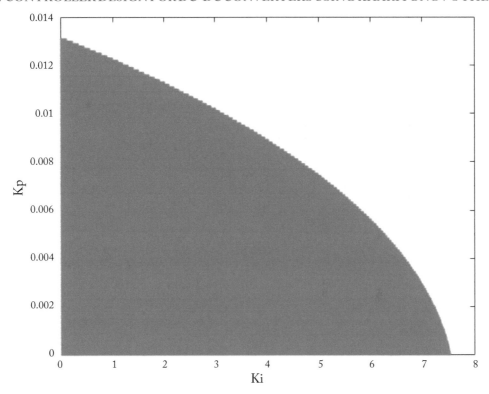

Figure 3.7: Acceptable PI coefficients (`kp_delta=.0001`).

The following scenario is used to test the close-loop system: input voltage's value changes from 42–54 V at $t = 50$ ms. Output load changes from 5–25 Ω at $t = 100$ ms and controller reference voltage changes from 12–20 V at $t = 150$ ms. The test scenario is summarized in Table 3.3.

Table 3.3: Test scenario for quadratic buck converter

Change In	Time	Initial Value	Final Value	$\dfrac{\text{Final} - \text{Initial}}{\text{Initial}} \times 100$
Vin	50 ms	42	54	28%
Rload	100 ms	5	25	400%
Vref	150 ms	12	20	66%

Selection of Kp (proportional gain) and Ki (integrator gain) are done using the obtained acceptable set (Fig. 3.7). For instance, we chose $Kp = 0$ and $Ki = 7$. The simulation result for the aforementioned scenario is shown in Fig. 3.8.

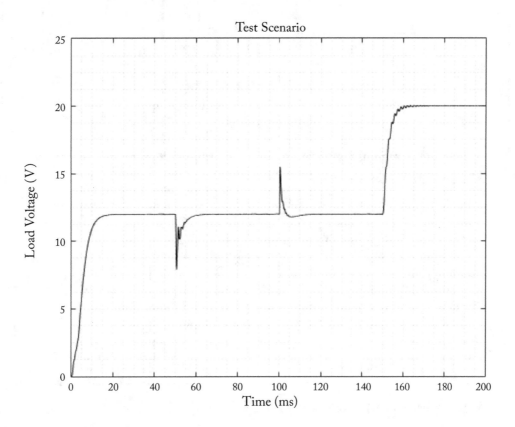

Figure 3.8: Simulation result.

You can test other acceptable values if obtained result does not meet specifications. Some trial error are required to find the best response.

3.3 ROBUST CONTROLLER DESIGN FOR QUADRATIC BOOST CONVERTER

Figure 3.9 shows the schematic of a quadratic boost converter.

When MOSFET M is closed, $D1$, $D2$, and $D3$ are reverse biased, forward biased, and reverse biased, respectively. Equivalent circuit for this case is shown in Fig. 3.10.

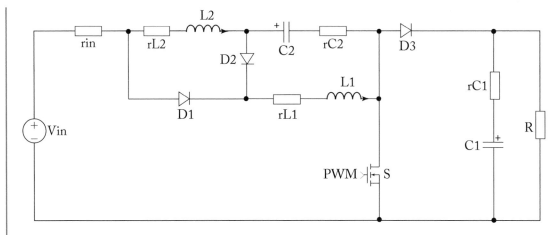

Figure 3.9: Quadratic boost converter.

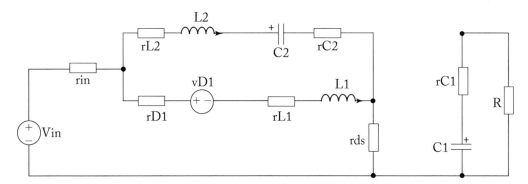

Figure 3.10: Equivalent circuit for closed MOSFET.

When MOSFET switch is closed, converters differential equation can be written as

$$
\begin{aligned}
L_1 \dot{i}_1 &= -(r_{D1} + r_{L1} + r_{in} + r_{ds})\, i_{L1} - (r_{in} + r_{ds})\, i_{L2} - v_{D1} + v_{in} \\
L_2 \dot{i}_2 &= -(r_{in} + r_{ds})\, i_{L1} - (r_{L2} + r_{C2} + r_{in} + r_{ds})\, i_{L2} - v_{C2} + v_{in} \\
C_1 \dot{v}_{C1} &= -\frac{v_{C1}}{R + r_{C1}} \\
C_2 \dot{v}_{C2} &= i_{L2}.
\end{aligned}
\tag{3.6}
$$

When MOSFET M is opened $D1$, $D2$, and $D3$ are forward biased, reverse biased, and reverse biased, respectively. Equivalent circuit for this case is shown in Fig. 3.11.

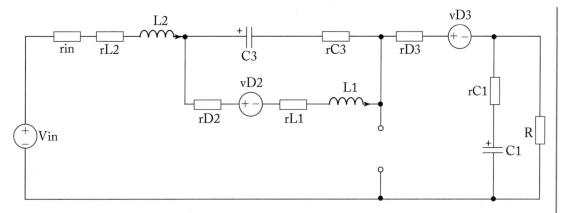

Figure 3.11: Equivalent circuit for opened MOSFET.

In this case converters dynamical equation can be written as:

$$L_1 \dot{i}_1 = -(r_{D2} + r_{L1} + r_{C2}) i_{L1} + r_{C2} i_{L2} + v_{C2} - v_{D2}$$

$$L_2 \dot{i}_2 = -\frac{R}{R + r_{C1}} v_{C1} - v_{C2} + v_{in} - v_{D3}$$

$$C_1 \dot{v}_{C1} = \frac{R}{R + r_{C1}} i_{L2} - \frac{v_{C1}}{R + r_{C1}}$$

$$C_2 \dot{v}_{C2} = i_{L2} - i_{L1}.$$

(3.7)

Assume a quadratic boost converter with the following parameter values: $V_{in} = 20$ V, $rin = 1\ \Omega$, $L1 = 500\ \mu H$, $rL1 = 20$ mΩ, $L2 = 500\ \mu H$, $rL2 = 20$ mΩ, $C1 = 100\ \mu F$, $rC1 = 25$ mΩ, $C2 = 22\ \mu F$, $rC2 = 150$ mΩ, $Fsw = 100$ KHz, $rD(on) = 10$ mΩ, $VD(on) = 0.8$ V, $rds = 100$ mΩ, and $48\ \Omega < R < 480\ \Omega$.

$rD(on)$, $VD(on)$ and rds shows diode forward bias resistance, diode forward voltage drop, and MOSFET on resistance, respectively. For the aforementioned load range converter operates in CCM.

Figure 3.12 shows the effect of changing output load resistance from 48–480 Ω (with 5 Ω steps) on the frequency response of converter.

The converter dynamical equation (control to output transfer function) can be written as:

$$\frac{\tilde{v}_o(s)}{\tilde{d}(s)} = \frac{n_3 s^3 + n_2 s^2 + n_1 s^1 + n_0}{d_4 s^4 + d_3 s^3 + d_2 s^2 + d_1 s^1 + d_0}.$$

(3.8)

The denominator is a fourth-order polynomial since there are four energy storage elements (two capacitors and two inductors) in the circuit.

The following MATLAB® code is used to extract the minimum and maximum of transfer function coefficients. Minimum and maximum of coefficients are shown in Table 3.4.

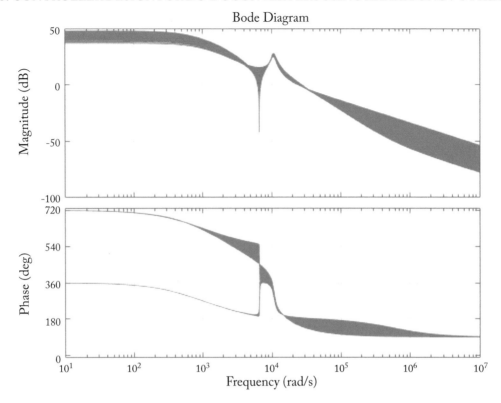

Figure 3.12: Control-to-output frequency response when the load changes from 48 Ω toward 480 Ω (with 5 Ω steps).

```
%This program extract the changes in coefficients of a
%Quadratic Boost Converter. 48<R<480
clc
clear all;

Rmin=48;
delta_R=5;
Rmax=480;

NUM=zeros(1,5);
DEN=zeros(1,5);
N=(Rmax-Rmin)/delta_R+2;
n=0;
```

```
for R=Rmin:delta_R:Rmax
n=n+1;
clc
disp('percentage of work done:');
disp(100*n/N)

D=.5; %Duty

VIN=20;
rin=1;

L1=500e-6;
rL1=.02;

L2=500e-6;
rL2=.02;

C1=100e-6;
rC1=.025;

C2=22e-6;
rC2=.15;

rD1=.01;
VD1=.8;
rD2=.01;
VD2=.8;
rD3=.01;
VD3=.8;

rds=.1;

%Symbolic equations
syms iL1 iL2 vC1 vC2 vin vD1 vD2 vD3 d

%CLOSED MOSFET EQUATIONS
x=(-(rds+rin)*(iL1+iL2)-(rL1+rD1)*iL1-vD1+vin);
M1=x/L1;
M2=(-(rL2+rC2)*iL2-vC2+(rD1+rL1)*iL1+vD1+x)/L2;
```

```
M3=-vC1/(R+rC1)/C1;
M4=iL2/C2;

%OPENED MOSFET EQUATIONS
y=iL2-iL1;
z=(iL2-(vC1/R))/(1+rC1/R);
t=-(rD2+rL1)*iL1-vD2+vC2+rC2*y;

M5=t/L1;
M6=(vin-(rin+rL2+rD3)*iL2-(rD2+rL1)*iL1-t-vD2-vD3-rC1*z-vC1)/L2;
M7=z/C1;
M8=y/C2;

%x=[iL1;iL2;vC1;vC2]

%AVERAGING
MA1= simplify(d*M1+(1-d)*M5);
MA2= simplify(d*M2+(1-d)*M6);
MA3= simplify(d*M3+(1-d)*M7);
MA4= simplify(d*M4+(1-d)*M8);

%DC OPERATING POINT CALCULATION
MA_DC_1=subs(MA1,[vin vD1 vD2 vD3 d],[VIN VD1 VD2 VD3 D]);
MA_DC_2=subs(MA2,[vin vD1 vD2 vD3 d],[VIN VD1 VD2 VD3 D]);
MA_DC_3=subs(MA3,[vin vD1 vD2 vD3 d],[VIN VD1 VD2 VD3 D]);
MA_DC_4=subs(MA4,[vin vD1 vD2 vD3 d],[VIN VD1 VD2 VD3 D]);

DC_SOL=
solve(MA_DC_1==0,MA_DC_2==0,MA_DC_3==0,MA_DC_4==0,
    'iL1','iL2','vC1','vC2');

IL1=eval(DC_SOL.iL1);
IL2=eval(DC_SOL.iL2);
VC1=eval(DC_SOL.vC1);
VC2=eval(DC_SOL.vC2);

%LINEARIZATION
%x=Ax+Bu
%x=[iL1;iL2;vC1;vC2]
```

```
%u=[vin;d] where d=duty and vin=small change in input voltage
A11=subs(simplify(diff(MA1,iL1)),
    [iL1 iL2 vC1 vC2 d vD1 vD2 vD3 vin],
    [IL1 IL2 VC1 VC2 D VD1 VD2 VD3 VIN]);
A12=subs(simplify(diff(MA1,iL2)),
    [iL1 iL2 vC1 vC2 d vD1 vD2 vD3 vin],
    [IL1 IL2 VC1 VC2 D VD1 VD2 VD3 VIN]);
A13=subs(simplify(diff(MA1,vC1)),
    [iL1 iL2 vC1 vC2 d vD1 vD2 vD3 vin],
    [IL1 IL2 VC1 VC2 D VD1 VD2 VD3 VIN]);
A14=subs(simplify(diff(MA1,vC2)),
    [iL1 iL2 vC1 vC2 d vD1 vD2 vD3 vin],
    [IL1 IL2 VC1 VC2 D VD1 VD2 VD3 VIN]);

A21=subs(simplify(diff(MA2,iL1)),
    [iL1 iL2 vC1 vC2 d vD1 vD2 vD3 vin],
    [IL1 IL2 VC1 VC2 D VD1 VD2 VD3 VIN]);
A22=subs(simplify(diff(MA2,iL2)),
    [iL1 iL2 vC1 vC2 d vD1 vD2 vD3 vin],
    [IL1 IL2 VC1 VC2 D VD1 VD2 VD3 VIN]);
A23=subs(simplify(diff(MA2,vC1)),
    [iL1 iL2 vC1 vC2 d vD1 vD2 vD3 vin],
    [IL1 IL2 VC1 VC2 D VD1 VD2 VD3 VIN]);
A24=subs(simplify(diff(MA2,vC2)),
    [iL1 iL2 vC1 vC2 d vD1 vD2 vD3 vin],
    [IL1 IL2 VC1 VC2 D VD1 VD2 VD3 VIN]);

A31=subs(simplify(diff(MA3,iL1)),
    [iL1 iL2 vC1 vC2 d vD1 vD2 vD3 vin],
    [IL1 IL2 VC1 VC2 D VD1 VD2 VD3 VIN]);
A32=subs(simplify(diff(MA3,iL2)),
    [iL1 iL2 vC1 vC2 d vD1 vD2 vD3 vin],
    [IL1 IL2 VC1 VC2 D VD1 VD2 VD3 VIN]);
A33=subs(simplify(diff(MA3,vC1)),
    [iL1 iL2 vC1 vC2 d vD1 vD2 vD3 vin],
    [IL1 IL2 VC1 VC2 D VD1 VD2 VD3 VIN]);
A34=subs(simplify(diff(MA3,vC2)),
    [iL1 iL2 vC1 vC2 d vD1 vD2 vD3 vin],
    [IL1 IL2 VC1 VC2 D VD1 VD2 VD3 VIN]);
```

```
A41=subs(simplify(diff(MA4,iL1)),
    [iL1 iL2 vC1 vC2 d vD1 vD2 vD3 vin],
    [IL1 IL2 VC1 VC2 D VD1 VD2 VD3 VIN]);
A42=subs(simplify(diff(MA4,iL2)),
    [iL1 iL2 vC1 vC2 d vD1 vD2 vD3 vin],
    [IL1 IL2 VC1 VC2 D VD1 VD2 VD3 VIN]);
A43=subs(simplify(diff(MA4,vC1)),
    [iL1 iL2 vC1 vC2 d vD1 vD2 vD3 vin],
    [IL1 IL2 VC1 VC2 D VD1 VD2 VD3 VIN]);
A44=subs(simplify(diff(MA4,vC2)),
    [iL1 iL2 vC1 vC2 d vD1 vD2 vD3 vin],
    [IL1 IL2 VC1 VC2 D VD1 VD2 VD3 VIN]);

A=eval([A11 A12 A13 A14;
        A21 A22 A23 A24;
        A31 A32 A33 A34;
        A41 A42 A43 A44]);

B11=subs(simplify(diff(MA1,vin)),
    [iL1 iL2 vC1 vC2 d vD1 vD2 vD3 vin],
    [IL1 IL2 VC1 VC2 D VD1 VD2 VD3 VIN]);
B12=subs(simplify(diff(MA1,d)),
    [iL1 iL2 vC1 vC2 d vD1 vD2 vD3 vin],
    [IL1 IL2 VC1 VC2 D VD1 VD2 VD3 VIN]);

B21=subs(simplify(diff(MA2,vin)),
    [iL1 iL2 vC1 vC2 d vD1 vD2 vD3 vin],
    [IL1 IL2 VC1 VC2 D VD1 VD2 VD3 VIN]);
B22=subs(simplify(diff(MA2,d)),
    [iL1 iL2 vC1 vC2 d vD1 vD2 vD3 vin],
    [IL1 IL2 VC1 VC2 D VD1 VD2 VD3 VIN]);

B31=subs(simplify(diff(MA3,vin)),
    [iL1 iL2 vC1 vC2 d vD1 vD2 vD3 vin],
    [IL1 IL2 VC1 VC2 D VD1 VD2 VD3 VIN]);
B32=subs(simplify(diff(MA3,d)),
    [iL1 iL2 vC1 vC2 d vD1 vD2 vD3 vin],
    [IL1 IL2 VC1 VC2 D VD1 VD2 VD3 VIN]);
```

```
B41=subs(simplify(diff(MA4,vin)),
    [iL1 iL2 vC1 vC2 d vD1 vD2 vD3 vin],
    [IL1 IL2 VC1 VC2 D VD1 VD2 VD3 VIN]);
B42=subs(simplify(diff(MA4,d)),
    [iL1 iL2 vC1 vC2 d vD1 vD2 vD3 vin],
    [IL1 IL2 VC1 VC2 D VD1 VD2 VD3 VIN]);

B=eval([B11 B12;
        B21 B22;
        B31 B32;
        B41 B42 ]);

yout=R/(R+rC1)*vC1*d+(iL2-z)*R*(1-d); %Averaged output
%equation Linearizing the averaged output equation
C_vR_1=subs(simplify(diff(yout,iL1)),
    [iL1 iL2 vC1 vC2 d vD1 vD2 vD3 vin],
    [IL1 IL2 VC1 VC2 D VD1 VD2 VD3 VIN]);
C_vR_2=subs(simplify(diff(yout,iL2)),
    [iL1 iL2 vC1 vC2 d vD1 vD2 vD3 vin],
    [IL1 IL2 VC1 VC2 D VD1 VD2 VD3 VIN]);
C_vR_3=subs(simplify(diff(yout,vC1)),
    [iL1 iL2 vC1 vC2 d vD1 vD2 vD3 vin],
    [IL1 IL2 VC1 VC2 D VD1 VD2 VD3 VIN]);
C_vR_4=subs(simplify(diff(yout,vC2)),
    [iL1 iL2 vC1 vC2 d vD1 vD2 vD3 vin],
    [IL1 IL2 VC1 VC2 D VD1 VD2 VD3 VIN]);

C_vR=eval([C_vR_1 C_vR_2 C_vR_3 C_vR_4]);

H_vR=tf(ss(A,B,C_vR,0));
vR_d=H_vR(1,2); %transfer function between duty ratio and
%output voltage(i.e., control-to-output transfer function)
%TF(:,:,n)=vR_d;
[num,den] = tfdata(vR_d,'v'); %extract the numerator and
%denominator of calculated transfer function
NUM=[NUM;num];
DEN=[DEN;den];
end
```

```
NUM(1,:)=[];
DEN(1,:)=[];

num_min=min(NUM); %minimum of numerator coefficients
num_max=max(NUM); %maximum pf numerator coefficients

den_min=min(DEN); %minimum of denominator coefficients
den_max=max(DEN); %maximum pf denominator coefficients
%End of code
```

Table 3.4: Minimum and maximum of control-to-output transfer function coefficients

Coefficient	Minimum	Maximum
n_3	-2.0219×10^4	-1.2205×10^3
n_2	3.8177×10^8	7.1462×10^8
n_1	-3.9968×10^{12}	5.8382×10^{10}
n_0	1.2446×10^{16}	3.1724×10^{16}
d_4	1	1
d_3	3.8059×10^3	3.9932×10^3
d_2	1.2105×10^8	1.2176×10^8
d_1	2.7068×10^{11}	2.9240×10^{11}
d_0	1.1910×10^{14}	1.6801×10^{14}

Assume a voltage control loop like that shown in Fig. 3.13. A PI controller is used to control the converter.

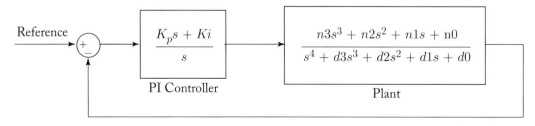

Figure 3.13: Structure of the control loop.

The close-loop transfer function is:

$$H(s) = \cfrac{\dfrac{K_p s + K_i}{s} \times \dfrac{n_3 s^3 + n_2 s^2 + n_1 s^1 + n_0}{s^4 + d_3 s^3 + d_2 s^2 + d_1 s^1 + d_0}}{1 + \dfrac{K_p s + K_i}{s} \times \dfrac{n_3 s^3 + n_2 s^2 + n_1 s^1 + n_0}{s^4 + d_3 s^3 + d_2 s^2 + d_1 s^1 + d_0}}. \tag{3.9}$$

After simplification the close-loop denominator obtained as:

$$s^5 + \left(d_3 + n_3 K_p\right) s^4 + \left(d_2 + n_2 K_p + n_3 K_i\right) s^3 + \left(d_1 + n_1 K_p + n_2 K_i\right) s^2$$
$$+ \left(d_0 + n_0 K_p + n_1 K_i\right) s + n_0 K_i. \tag{3.10}$$

Minimum and maximum of these coefficients are shown in Table 3.5.

Table 3.5: Minimum and maximum of close-loop coefficients (Equation (3.10))

Coefficient	Minimum	Maximum
$d_3 + n_3 K_p$	$3.8059 \times 10^3 - 2.0219 \times 10^4 K_p$	$3.9932 \times 10^3 - 1.2205 \times 10^3 K_p$
$d_2 + n_2 K_p + n_3 K_i$	$1.2105 \times 10^8 + 3.8177 \times 10^8 K_p$ $- 2.0219 \times 10^4 K_i$	$1.2176 \times 10^8 + 7.1462 \times 10^8 K_p$ $- 1.2205 \times 10^3 K_i$
$d_1 + n_1 K_p + n_2 K_i$	$2.7068 \times 10^{11} - 3.9968 \times$ $10^{12} K_p + 3.8177 \times 10^8 K_i$	$2.9240 \times 10^{11} + 5.8382 \times$ $10^{10} K_p + 7.1462 \times 10^8 K_i$
$d_0 + n_0 K_p + n_1 K_i$	$1.1910 \times 10^{14} + 1.2446 \times 10^{16} K_p$ $- 3.9968 \times 10^{12} K_i$	$1.6801 \times 10^{14} + 3.1724 \times$ $10^{16} K_p + 5.8382 \times 10^{10} K_i$
$n_0 K_i$	$1.2446 \times 10^{16} K_i$	$3.1724 \times 10^{16} K_i$

We use this table to form the Kharitonov's polynomials. Since the closed-loop denominator is a fifth-order polynomial, we need to stabilize only $K_2(s)$, $K_3(s)$, and $K_4(s)$ (see Lemma 2.8).

The following code finds the suitable values of Kp and Ki which stabilize the $K_2(s)$, $K_3(s)$, and $K_4(s)$. Acceptable region found by the code is shown in Fig. 3.14.

```
clc

kp_min=0;
kp_max=.06;
kp_delta=.001;

ki_min=0;
```

```matlab
ki_max=30;
ki_delta=.01;

sol=[0 0]; %sol=[kp ki]

N=(kp_max-kp_min)/kp_delta*(ki_max-ki_min)/ki_delta;
n=0;

for kp=kp_min:kp_delta:kp_max
    for ki=ki_min:ki_delta:ki_max

        n=n+1;
        disp('percentage of work done')
        100*n/N
        disp('---')
        %                        5     4     3     2     1
        %denominator is :s +z4*s +z3*s +z2*s +z1*s +z0
        %maxsimum is shown with p
        %minimum is shown with n
        %For example, z4_p shows the z4 maximum.
        %z4 is s^4 coefficient.
        %z4_n shows the z4 minimum. z4 is s^4 coefficient.

        z4_p=3.9932e3-1.2205e3*kp;
        z4_n=3.8059e3-20.219e3*kp;


        z3_p=1.2176e8+7.1462e8*kp-1.2205e3*ki;
        z3_n=1.2105e8+3.8177e8*kp-20.219e3*ki;


        z2_p=2.924e11+5.8382e10*kp+7.1462e8*ki;
        z2_n=2.7065e11-3.9968e12*kp+3.8177e8*ki;


        z1_p=1.6801e14+3.1724e16*kp+5.8382e10*ki;
        z1_n=1.191e14+1.2446e16*kp-3.9968e12*ki;


        z0_p=3.1724e16*ki;
        z0_n=1.2446e16*ki;

        %Since the interval polynomial is of 5th order
```

```
%stabilization of K2(s), K3(s), and K4(s) is enough.
%See Lemma 2.1

    k2=roots([1 z4_p z3_n z2_n z1_p z0_p]); %K2(s)
    k2_real=real(k2);
    T2=sum(k2_real>0); %number of unstable roots in K2(s)

    k3=roots([1 z4_n z3_p z2_n z1_p z0_n]);
    k3_real=real(k3);
    T3=sum(k3_real>0); %number of unstable roots in K3(s)

    k4=roots([1 z4_p z3_n z2_p z1_n z0_p]);
    k4_real=real(k4);
    T4=sum(k4_real>0); %number of unstable roots in K4(s)

    %When T2+T3+T4 equals to zero K2(s),
    %K3(s) and K4(s) are Hurwitz.
    %The associated kp and ki are acceptable.
    if ((T2+T3+T4)==0)
        sol=[sol;[kp ki]];
    end
    end
end

sol(1,:)=[]; %Removes the initialization([0 0])

plot(sol(:,2),sol(:,1),'.'), xlabel('Ki'), ylabel ('Kp'),
    grid minor
%End of code
```

The following scenario is used to test the close-loop system: output load changes from 48–480 Ω at $t = 50$ ms, controller reference voltage changes from 30–45 V at $t = 100$ ms, and input voltage value changes from 20–16 V at $t = 150$ ms. The test scenario is summarized in Table 3.6.

Selection of *Kp* (proportional gain) and *Ki* (integrator gain) are done using the obtained acceptable set (Fig. 3.14). For instance, we chose, *Kp* = 0 and *Ki* = 2.8. The simulation result for the aforementioned scenario is shown in Fig. 3.15. You can test other values in the acceptable region.

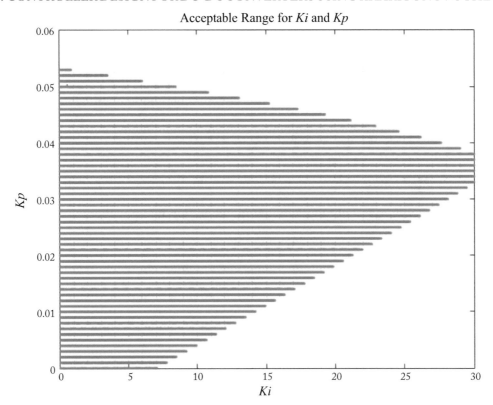

Figure 3.14: Acceptable Kp and Ki.

Table 3.6: Test scenario for quadratic boost converter

Change In	Time	Initial Value	Final Value	$\dfrac{\text{Final} - \text{Initial}}{\text{Initial}} \times 100$
Rload	50 ms	48	480	+900%
Vref	100 ms	30	45	+50%
Vin	150 ms	20	16	− 20%

3.4   ROBUST CONTROLLER DESIGN FOR SUPER BUCK CONVERTER

Figure 3.16 shows the schematic of a super buck converter.

When MOSFET M is closed diode $D1$ is reverse biased. Figure 3.17 shows the equivalent circuit for this case.

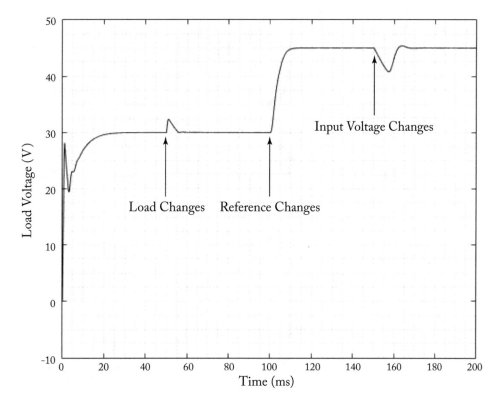

Figure 3.15: Simulation result.

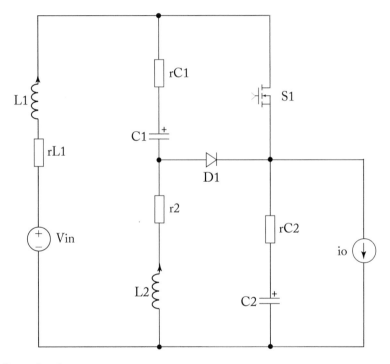

Figure 3.16: **Super buck converter.**

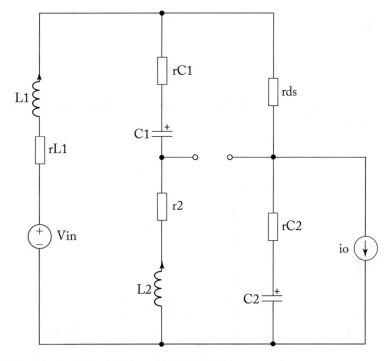

Figure 3.17: Equivalent circuit for closed MOSFET.

When MOSFET switch is closed, converter differential equations can be written as

$$
\dot{i}_{L1} = -\frac{r_{L1} + r_{ds} + r_{C2}}{L_1}i_{L1} - \frac{r_{ds} + r_{C2}}{L_1}i_{L2} - \frac{v_{C2}}{L_1} + \frac{v_{in}}{L_1} + \frac{r_{C2}}{L_1}i_o
$$

$$
\dot{i}_{L2} = -\frac{r_{ds} + r_{C2}}{L_2}i_{L1} - \frac{r_{L2} + r_{ds} + r_{C1} + r_{C2}}{L_2}i_{L2} + \frac{v_{C1}}{L_2} - \frac{v_{C2}}{L_2} + \frac{r_{C2}}{L_2}i_o
$$

$$
\dot{v}_{C1} = -\frac{i_{L2}}{C_1}
$$

$$
\dot{v}_{C2} = \frac{i_{L1}}{C_2} + \frac{i_{L2}}{C_2} - \frac{i_o}{C_2}.
$$

(3.11)

When MOSFET M is opened, diode $D1$ is forward biased. Figure 3.18 shows the equivalent circuit for this case.

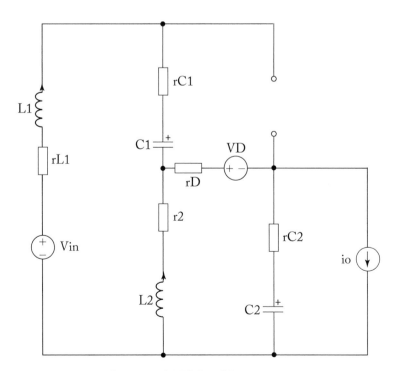

Figure 3.18: Equivalent circuit for opened MOSFET.

In this case converter dynamical equations can be written as:

$$\dot{i_{L1}} = -\frac{r_{L1} + r_{C1} + r_d + r_{C2}}{L_1}i_{L1} - \frac{r_D + r_{C2}}{L_1}i_{L2}$$

$$-\frac{v_{C1}}{L_1} - \frac{v_{C2}}{L_1} + \frac{v_{in}}{L_1} + \frac{r_{C2}}{L_1}i_o - \frac{V_D}{L_1}$$

$$\dot{i_{L2}} = -\frac{r_D + r_{C2}}{L_2}i_{L1} - \frac{r_{L2} + r_D + r_{C2}}{L_2}i_{L2} - \frac{v_{C2}}{L_2} + \frac{r_{C2}}{L_2}i_o - \frac{V_D}{L_2} \qquad (3.12)$$

$$\dot{v_{C1}} = \frac{i_{L1}}{C_1}$$

$$\dot{v_{C2}} = \frac{i_{L1}}{C_2} + \frac{i_{L2}}{C_2} - \frac{i_o}{C_2}.$$

Assume a super buck converter with the following parameter values: $V_{in} = 30$ V, $L1 = 500\,\mu$H, $r_{L1} = 0.1\,\Omega$, $L2 = 500\,\mu$H, $r_{L2} = 0.1\,\Omega$, $C1 = 470\,\mu$F, $r_{C1} = 0.05\,\Omega$, $C2 = 470\,\mu$F, $r_{C2} = 0.05\,\Omega$, $5\,\Omega < R < 25\,\Omega$, $D = 0.66$, and $Fsw = 50$ KHz.

D and Fsw show the duty ratio and the switching frequency of the converter, respectively. Figure 3.19 shows the frequency response of control-to-output transfer function when load resistor changes from 5 Ω toward 25 Ω with 0.2 Ω steps. For the aforementioned load range the converter operates in the CCM.

The converter dynamical equation (control to output transfer function) can be written as:

$$\frac{\tilde{v}_o(s)}{\tilde{d}(s)} = \frac{n_3 s^3 + n_2 s^2 + n_1 s^1 + n_0}{s^4 + d_3 s^3 + d_2 s^2 + d_1 s^1 + d_0}. \qquad (3.13)$$

The denominator is a fourth-order polynomial since there are four energy storage elements (two capacitors and two inductors) in the circuit.

The following MATLAB® code is used to extract the minimum and maximum of transfer function coefficients. Minimum and maximum of coefficients is shown in Table 3.7.

```
clc
clear all;

%R=5..20

R_min=5;
R_step=.2;
R_max=25;

NUM=zeros(1,5);
DEN=zeros(1,5);
```

```
N=(R_max-R_min)/R_step;
n=0;

for R=R_min:R_step:R_max
n=n+1;
disp('percentage of work done:')
disp(n/N*100)

VIN=30;
rin=.1;

L1=500e-6;
rL1=.1;

L2=500e-6;
rL2=.1;

C1=470e-6;
rC1=.05;

C2=470e-6;
rC2=.05;

rD=.01;
VD=.7;

rds=.1;

D=.66;

%Symbolic equations
syms iL1 iL2 vC1 vC2 vin vD d

%CLOSED MOSFET EQUATIONS
y=(iL1+iL2-(vC2/R))/(1+(rC2/R));
x=(rC1+rL2)*iL2-vC1+rds*(iL1+iL2)+rC2*y+vC2;
M1=(vin-x-vC1+(rL2+rC1)*iL2-rL1*iL1)/L1;
M2=(-(rC1+rL2)*iL2+vC1-rds*(iL1+iL2)-rC2*y-vC2)/L2;
```

```
M3=(-iL2)/C1;
M4=(y)/C2;

%OPENED MOSFET EQUATIONS
z=vD+rD*(iL1+iL2)+rC2*y+vC2+rL2*iL2;
M5=(vin-(rL1+rC1)*iL1-vC1+rL2*iL2-z)/L1;
M6=(-vD-rD*(iL1+iL2)-rC2*y-vC2-rL2*iL2)/L2;
M7=(iL1)/C1;
M8=(y)/C2;

%AVERAGING
MA1= simplify(d*M1+(1-d)*M5);
MA2= simplify(d*M2+(1-d)*M6);
MA3= simplify(d*M3+(1-d)*M7);
MA4= simplify(d*M4+(1-d)*M8);

%DC OPERATING POINT CALCULATION
MA_DC_1=subs(MA1,[vin vD d],[VIN VD D]);
MA_DC_2=subs(MA2,[vin vD d],[VIN VD D]);
MA_DC_3=subs(MA3,[vin vD d],[VIN VD D]);
MA_DC_4=subs(MA4,[vin vD d],[VIN VD D]);

DC_SOL=
solve(MA_DC_1==0,MA_DC_2==0,MA_DC_3==0,MA_DC_4==0,
    'iL1','iL2','vC1','vC2');

IL1=eval(DC_SOL.iL1);
IL2=eval(DC_SOL.iL2);
VC1=eval(DC_SOL.vC1);
VC2=eval(DC_SOL.vC2);

%LINEARIZATION
%x=[iL1;iL2;vC1;vC2]
%u=[vin;d] where d=duty and vin=input voltage changes
A11=subs(simplify(diff(MA1,iL1)),[iL1 iL2 vC1 vC2 d vD vin],
    [IL1 IL2 VC1 VC2 D VD VIN]);
A12=subs(simplify(diff(MA1,iL2)),[iL1 iL2 vC1 vC2 d vD vin],
    [IL1 IL2 VC1 VC2 D VD VIN]);
A13=subs(simplify(diff(MA1,vC1)),[iL1 iL2 vC1 vC2 d vD vin],
```

```
     [IL1 IL2 VC1 VC2 D VD VIN]);
A14=subs(simplify(diff(MA1,vC2)),[iL1 iL2 vC1 vC2 d vD vin],
     [IL1 IL2 VC1 VC2 D VD VIN]);


A21=subs(simplify(diff(MA2,iL1)),[iL1 iL2 vC1 vC2 d vD vin],
     [IL1 IL2 VC1 VC2 D VD VIN]);
A22=subs(simplify(diff(MA2,iL2)),[iL1 iL2 vC1 vC2 d vD vin],
     [IL1 IL2 VC1 VC2 D VD VIN]);
A23=subs(simplify(diff(MA2,vC1)),[iL1 iL2 vC1 vC2 d vD vin],
     [IL1 IL2 VC1 VC2 D VD VIN]);
A24=subs(simplify(diff(MA2,vC2)),[iL1 iL2 vC1 vC2 d vD vin],
     [IL1 IL2 VC1 VC2 D VD VIN]);


A31=subs(simplify(diff(MA3,iL1)),[iL1 iL2 vC1 vC2 d vD vin],
     [IL1 IL2 VC1 VC2 D VD VIN]);
A32=subs(simplify(diff(MA3,iL2)),[iL1 iL2 vC1 vC2 d vD vin],
     [IL1 IL2 VC1 VC2 D VD VIN]);
A33=subs(simplify(diff(MA3,vC1)),[iL1 iL2 vC1 vC2 d vD vin],
     [IL1 IL2 VC1 VC2 D VD VIN]);
A34=subs(simplify(diff(MA3,vC2)),[iL1 iL2 vC1 vC2 d vD vin],
     [IL1 IL2 VC1 VC2 D VD VIN]);


A41=subs(simplify(diff(MA4,iL1)),[iL1 iL2 vC1 vC2 d vD vin],
     [IL1 IL2 VC1 VC2 D VD VIN]);
A42=subs(simplify(diff(MA4,iL2)),[iL1 iL2 vC1 vC2 d vD vin],
     [IL1 IL2 VC1 VC2 D VD VIN]);
A43=subs(simplify(diff(MA4,vC1)),[iL1 iL2 vC1 vC2 d vD vin],
     [IL1 IL2 VC1 VC2 D VD VIN]);
A44=subs(simplify(diff(MA4,vC2)),[iL1 iL2 vC1 vC2 d vD vin],
     [IL1 IL2 VC1 VC2 D VD VIN]);


A=eval([A11 A12 A13 A14;
        A21 A22 A23 A24;
        A31 A32 A33 A34;
        A41 A42 A43 A44
        ]);


B11=subs(simplify(diff(MA1,vin)),[iL1 iL2 vC1 vC2 d vD vin],
     [IL1 IL2 VC1 VC2 D VD VIN]);
```

```
B12=subs(simplify(diff(MA1,d)),[iL1 iL2 vC1 vC2 d vD vin],
    [IL1 IL2 VC1 VC2 D VD VIN]);

B21=subs(simplify(diff(MA2,vin)),[iL1 iL2 vC1 vC2 d vD vin],
    [IL1 IL2 VC1 VC2 D VD VIN]);
B22=subs(simplify(diff(MA2,d)),[iL1 iL2 vC1 vC2 d vD vin],
    [IL1 IL2 VC1 VC2 D VD VIN]);

B31=subs(simplify(diff(MA3,vin)),[iL1 iL2 vC1 vC2 d vD vin],
    [IL1 IL2 VC1 VC2 D VD VIN]);
B32=subs(simplify(diff(MA3,d)),[iL1 iL2 vC1 vC2 d vD vin],
    [IL1 IL2 VC1 VC2 D VD VIN]);

B41=subs(simplify(diff(MA4,vin)),[iL1 iL2 vC1 vC2 d vD vin],
    [IL1 IL2 VC1 VC2 D VD VIN]);
B42=subs(simplify(diff(MA4,d)),[iL1 iL2 vC1 vC2 d vD vin],
    [IL1 IL2 VC1 VC2 D VD VIN]);

B=eval([B11 B12;
        B21 B22;
        B31 B32;
        B41 B42
        ]);

C_vR=R*[1-(1/(1+rC2/R)) 1-(1/(1+rC2/R)) 0 1/R/(1+(rC2/R))];

H_vR=tf(ss(A,B,C_vR,0));
vR_d=H_vR(1,2);

[num,den] = tfdata(vR_d,'v');
NUM=[NUM;num];
DEN=[DEN;den];
end

NUM(1,:)=[];
DEN(1,:)=[];

num_min=min(NUM); %minimum of numerator coefficients
num_max=max(NUM); %maximum pf numerator coefficients
```

```
den_min=min(DEN); %minimum of denominator coefficients
den_max=max(DEN); %maximum pf denominator coefficients
```

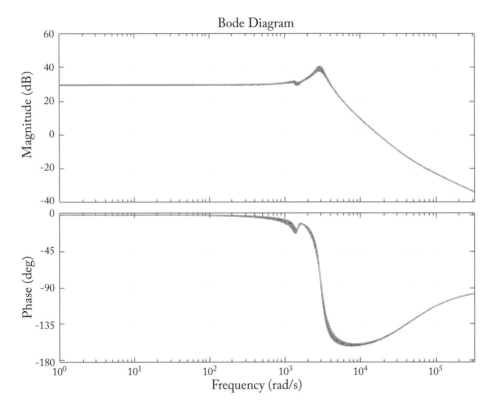

Figure 3.19: Change in control-to-output frequency response when load changes from 5–25 Ω.

Assume a voltage control loop like that shown in Fig. 3.20. A PI controller is used to control the converter.

The close-loop transfer function is:

$$H(s) = \frac{\dfrac{K_p s + K_i}{s} \times \dfrac{n_3 s^3 + n_2 s^2 + n_1 s^1 + n_0}{s^4 + d_3 s^3 + d_2 s^2 + d_1 s^1 + d_0}}{1 + \dfrac{K_p s + K_i}{s} \times \dfrac{n_3 s^3 + n_2 s^2 + n_1 s^1 + n_0}{s^4 + d_3 s^3 + d_2 s^2 + d_1 s^1 + d_0}}. \tag{3.14}$$

Table 3.7: Minimum and maximum of control-to-output transfer function coefficients

Coefficient	Minimum	Maximum
n_3	5.9932×10^3	6.1026×10^3
n_2	2.5627×10^8	2.6114×10^8
n_1	6.5557×10^{10}	7.5091×10^{10}
n_0	5.4098×10^{14}	5.521×10^{14}
d_4	1	1
d_3	1.0832×10^3	1.3969×10^3
d_2	1.1099×10^7	1.1281×10^7
d_1	3.9649×10^9	4.7268×10^9
d_0	1.8185×10^{13}	1.8415×10^{13}

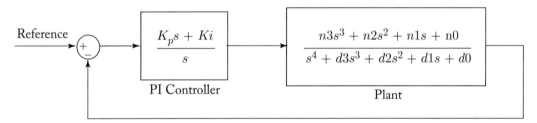

Figure 3.20: Structure of control loop.

After simplification the close-loop denominator is:

$$s^5 + \left(d_3 + n_3 K_p\right) s^4 + \left(d_2 + n_2 K_p + n_3 K_i\right) s^3 + \left(d_1 + n_1 K_p + n_2 K_i\right) s^2 + \left(d_0 + n_0 K_p + n_1 K_i\right) s + n_0 K_i. \tag{3.15}$$

Minimum and maximum of these coefficients are shown in Table 3.8.

We use this table to form Kharitonov's polynomials. Since the closed-loop denominator is a fifth-order polynomial, we need to stabilize only $K_2(s)$, $K_3(s)$, and $K_4(s)$ (see Lemma 2.8).

The following code finds the suitable values of Kp and Ki which stabilize the $K_2(s)$, $K_3(s)$, and $K_4(s)$. Acceptable region found by the code is shown in Fig. 3.21. In this case nearly all the points are acceptable.

```
clc

kp_min=0;
kp_max=20;
kp_delta=.1;

ki_min=0;
ki_max=20;
ki_delta=.1;

sol=[0 0]; %sol=[kp ki]

N=(kp_max-kp_min)/kp_delta*(ki_max-ki_min)/ki_delta;
n=0;

jevab=0;
for kp=kp_min:kp_delta:kp_max
    for ki=ki_min:ki_delta:ki_max

        n=n+1;
        disp('percentage of work done')
        100*n/N
        disp('---')
        %                 5      4      3      2      1
        %denominator is :s +z4*s +z3*s +z2*s +z1*s +z0
        %maxsimum is shown with p
        %minimum is shown with n
        %For example, z4_p shows the z4 maximum.
        %z4 is s^4 coefficient.
        %z4_n shows the z4 minimum. z4 is s^4 coefficient.

        z4_p=1.3969e3+6.1026e3*kp;
        z4_n=1.0832e3+5.9932e3*kp;

        z3_p=1.1281e7+2.6114e8*kp+6.1026e3*ki;
        z3_n=1.1099e7+2.5627e8*kp+5.9932e3*ki;

        z2_p=4.7268e9+7.5091e10*kp+2.6114e8*ki;
```

```
z2_n=3.9649e9+6.5557e10*kp+2.5627e8*ki;

z1_p=1.8415e13+5.521e14*kp+7.5091e10*ki;
z1_n=1.8185e13+5.4098e14*kp+6.5557e10*ki;

z0_p=5.521e14*ki;
z0_n=5.4098e14*ki;

%Since the interval polynomial is of 5th order
%stabilization of K2(s), K3(s) and K4(s) is enough.
%See Lemma 2.1

 k2=roots([1 z4_p z3_n z2_n z1_p z0_p]);
 k2_real=real(k2);
 T2=sum(k2_real>0); %number of unstable roots in K2(s)

 k3=roots([1 z4_n z3_p z2_n z1_p z0_n]);
 k3_real=real(k3);
 T3=sum(k3_real>0); %number of unstable roots in K3(s)

 k4=roots([1 z4_p z3_n z2_p z1_n z0_p]);
 k4_real=real(k4);
 T4=sum(k4_real>0); %number of unstable roots in K4(s)

 %When T2+T3+T4 equals to zero K2(s),
 %K3(s) and K4(s) are Hurwitz.
 %The associated kp and ki are acceptable.
 if ((T2+T3+T4)==0)
     sol=[sol;[kp ki]];
 end
     end
end
end

sol(1,:)=[]; %Removes the initialization([0 0])

plot(sol(:,2),sol(:,1),'.'), xlabel('Ki'), ylabel ('Kp'),
    grid minor
```

Table 3.8: Minimum and maximum of close-loop coefficients (Equation (3.15))

Coefficient	Minimum	Maximum
$d_3 + n_3 K_p$	$1.0832 \times 10^3 + 5.9932 \times 10^3 K_p$	$1.3969 \times 10^3 + 6.1026 \times 10^3 K_p$
$d_2 + n_2 K_p + n_3 K_i$	$1.1099 \times 10^7 + 2.5627 \times 10^8 K_p$ $+ 5.9932 \times 10^3 K_i$	$1.1281 \times 10^7 + 2.6114 \times 10^8 K_p$ $+ 6.1026 \times 10^3 K_i$
$d_1 + n_1 K_p + n_2 K_i$	$3.9649 \times 10^9 + 6.5557 \times 10^{10} K_p$ $+ 2.5627 \times 10^8 K_i$	$4.7268 \times 10^9 + 7.5091 \times 10^{10} K_p$ $+ 6.6114 \times 10^8 K_i$
$d_0 + n_0 K_p + n_1 K_i$	$1.8185 \times 10^{13} + 5.4098 \times 10^{14} K_p$ $+ 6.5557 \times 10^{10} K_i$	$1.8415 \times 10^{13} + 5.521 \times 10^{14} K_p$ $+ 7.5091 \times 10^{10} K_i$
$n_0 K_i$	$5.4098 \times 10^{14} K_i$	$5.521 \times 10^{14} K_i$

The following scenario is used to test the close-loop system: output load changes from 5–25 Ω at $t = 100$ ms, input voltage's value changes from 30–40 V at $t = 150$ ms, and controller reference voltage changes from 20–15 V at $t = 250$ ms. Test scenario is summarized in Table 3.9.

Table 3.9: Test scenario for quadratic buck converter

Change In	Time	Initial Value	Final Value	$\dfrac{\text{Final} - \text{Initial}}{\text{Initial}} \times 100$
Rload	100 ms	5	20	+300%
Vin	150 ms	30	40	+28%
Vref	250 ms	20	15	− 25%

Selection of *Kp* (proportional gain) and *Ki* (integrator gain) are done using the obtained acceptable set (Fig. 3.21). For instance, we chose, *Kp* = 0.2 and *Ki* = 9.7. The simulation result for the aforementioned scenario is shown in Fig. 3.22. Figures 3.23–3.25 show a closer view of Fig. 3.23. You can test other values in the acceptable region.

3.5 CONCLUSION

In this chapter we designed controllers for DC-DC converters using Kharitonov's theorem. We showed the design procedure with the aid of three examples. The proposed method can be applied to other DC-DC converters as well.

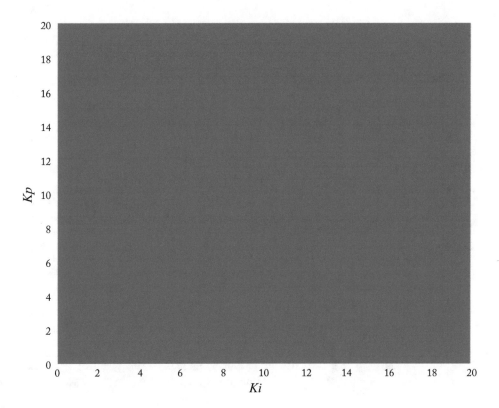

Figure 3.21: Acceptable *Kp* and *Ki*.

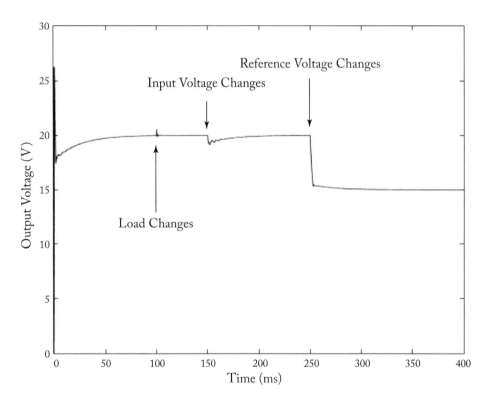

Figure 3.22: Simulation result.

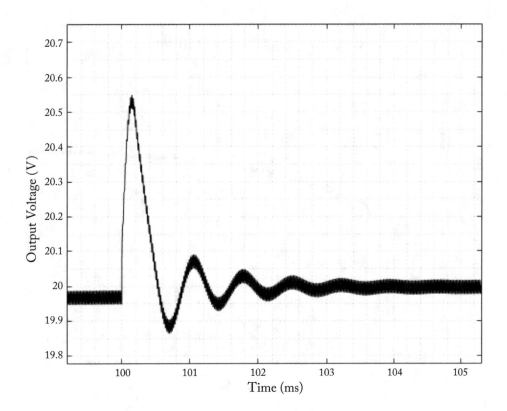

Figure 3.23: Output voltage changes due to load changes.

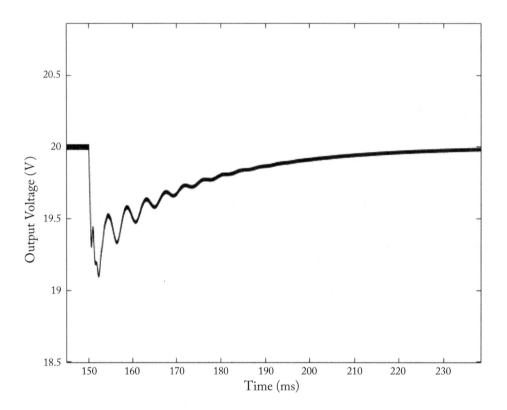

Figure 3.24: Output voltage changes due to input source changes.

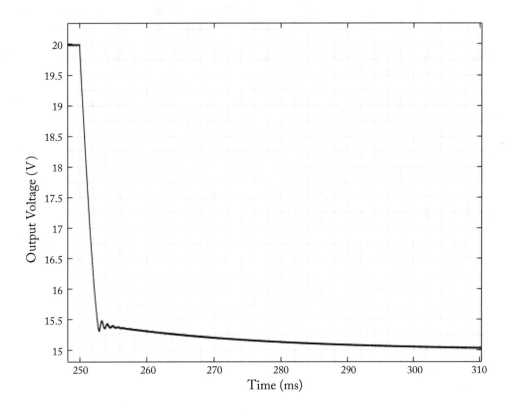

Figure 3.25: Tracking a different reference voltage.

REFERENCES

Ang, S. and Oliva, A. *Power Switching Converters*, Taylor & Francis, 2005.

Bacha, S., Munteanu, I., and Bratcu, A. I. *Power Electronics Converters Modeling and Control*, Springer, 2014. DOI: 10.1007/978-1-4471-5478-5.

Belanger, P. R. *Control Engineering: A Modern Approach*, Oxford University Press, 2005.

Bhattacharyya, S. P., Chapellat, H., and Keel, L. H. *Robust Control: The Parametric Approach*, Prentice Hall, 1995. DOI: 10.1016/b978-0-08-042230-5.50016-5.

Kazimierczuck, M. K. *Pulse Width Modulated DC-DC Power Converters*, John Wiley, 2012. DOI: 10.1002/9780470694640.

Ross Barmish, B. *New Tools for Robustness of Linear Systems*, Macmillan Publishing Company, 1994.

Sira Ramirez, H. and Silva Ortigoza, R. *Control Design Techniques in Power Electronics Devices*, Springer, 2006. DOI: 10.1007/1-84628-459-7.

Suntio, T. *Dynamic Profile of Switched Mode Converter: Modeling, Analysis and Control*, Wiley VCH, 2009. DOI: 10.1002/9783527626014.

Vasca, F. and Iannelli, L. *Dynamics and Control of Switched Electronic Systems*, Springer, 2012. DOI: 10.1007/978-1-4471-2885-4.

Author's Biography

FARZIN ASADI

Farzin Asadi received his B.Sc. in Electronics Engineering, his M.Sc. degree in Control Engineering, and his Ph.D. in Mechatronics Engineering. Currently, he is with the Department of Mechatronics Engineering at the Kocaeli University, Kocaeli, Turkey.

Farzin has published 25 international papers and 6 books. He is on the editorial board of 6 scientific journals as well. His research interests include switching converters, control theory, robust control of power electronics converters, and robotics.